D0892074

KEEPING BUILDINGS HEALTHY

KEEPING BUILDINGS HEALTHY

How to Monitor and Prevent
Indoor Environmental Problems

James T. O'Reilly
Philip Hagan
Ronald Gots
Alan Hedge

A WILEY-INTERSCIENCE PUBLICATION
JOHN WILEY & SONS, INC.

New York ▪ Chichester ▪ Weinheim ▪ Brisbane ▪ Singapore ▪ Toronto

Library of Congress Cataloging-in-Publication Data:

Keeping buildings healthy : how to monitor and prevent indoor
    environmental problems / James T. O'Reilly ... [et al.].
            p.   cm.
        Includes index.
        ISBN 0-471-29228-1 (cloth : alk. paper)
        1. Sick building syndrome.   2. Indoor air pollution.
    I. O'Reilly, James T., 1947–   .
    RA566.6.K44   1998
    613'.5—dc21   98-3365
                                                    CIP

Printed in the United States of America.

10 9 8 7 6 5 4 3 2

*This book is dedicated
to Maureen O'Reilly
with thanks for her
laughter, love, and joy*

Contents

Introduction

Building managers and owners have the difficult task of dealing effectively with claims of illness related to the condition of their buildings. The indoor environmental health issue is a complex matrix of medical, psychological, and engineering issues.

This book acknowledges that *some* building environments cause *some* persons to become ill. There is no point in denying that possibility as a scientific, medical, or legal matter. But this book's message to the building manager and owner is that building illness allegations need to be dealt with in the context of *people* problems. Science alone is not sufficient.

Complaining about work is a human trait. People normally complain about their working conditions at varying rates of frequency and intensity. When the complaints concern building conditions as a cause of physical harm, a great deal of patience and selectivity is needed. Not all complaints will be valid, but not all are exaggerated. Professional help is essential. Deciding which health complaints about building conditions should be given medical and scientific acceptance requires a lot of experience. For employee and tenant complaints, the building manager's skills in listening and responding can be the most cost-effective measures. Empathy in listening is the first step; selecting the right technical and medical follow-up comes later.

This book addresses another real issue surrounding health complaints by building occupants: cost control. Allegations not handled well today may ripen into lawsuits or compensation conflicts tomorrow. Staying out of the costly legal system by spending money up front on workers' health protection makes great sense. Expensive risks to the profitability of the building owner can be lessened by adequate planning and a thorough understanding of the technical and medical issues involved with sick building syndrome (SBS).

The unique collaboration this book represents supplies a tool kit for the building manager or owner. As with any tool kit, there is no magic formula for when a do-it-yourself project should be turned over to a more experienced worker who specializes in this type of response. The book's contributors are medical, psychological, legal, and technical experts whose work in indoor health issues enables them to guide the reader through the basic steps of effective response to SBS allegations. We seek to educate and prepare the person who will provide the first line of response to alleged health risks and guide the progress of the follow-up so that the building owner can be assured that the best course of action is taken.

The following list summarizes the main points the reader can expect to encounter in this book.

1. The number of allegations of illness from indoor air environmental exposures have apparently increased, though there are no universal statistical collections. Many more allegations of building-caused illnesses arise than can be verified by medical or scientific proof. Each case needs to be separately examined, but the majority of illness allegations will not be susceptible of proof. Empathetic listening is a core value for the resolution of these allegations.

2. Medical evaluation of the complaining person should be done with close attention to the results of on-site industrial hygiene measurements. The diagnostic uncertainties of the evaluation process may be daunting for the physician who has not previously been experienced in this form of evaluation. Knowing what to look for, what recognition of causes is possible, and what to rule out will be essential to successful medical responses.

3. The psychological consequences upon other employees of a complaining employee's attribution of their illness to causes in the building should not be overlooked. Group responses to stimuli and suggestion are important aspects of the several major building evacuation examples, addressed in the chapter on psychological issues. The prudent building manager will address the individual's complaint rapidly and factually and will be ready if there is a further allegation that others have developed the same symptoms.

4. Technical solutions to building discomfort are not difficult, in many cases; the engineer who works with these kinds of problems is both technician and ambassador/counselor to the alarmed employees. The engineer will need to understand about ventilation rates, air flow, some aspects of microbiology, and something about the interpersonal skills of alleviating concern among persons fearful of the unknown. The chapters of this text dealing with indoor air quality (IAQ) problems and solutions offer practical suggestions for the use of this technical intermediary.

5. Lawyers can play a constructive or destructive role in the response to sick building complaints. It is the client who dictates an accommodation strategy or who demands to fight to the last bullet against any assertion of building problems. Lawyers can be constructive when they avoid ill-advised terminations of a complaining worker who has a legitimate complaint and when they guide negotiations between the insurance carrier and the building owner. Too often, lawyers tend to be oriented to long, expensive and adversarial processes. Clients, includ-

ing building managers, need to be aware of the decisions that their lawyers are making and how their direction will affect what the lawyer will deliver.

6. In general, the destructive aspects of legal responses to sick-building allegations come from the incentive of contingent fee tort lawsuits, which induce plaintiff lawyers to sign up groups of building-resident employees to assert claims of injury. The claims can then be settled as a way to pay off the challenger without the cost of a trial; taking the case all the way through verdict and appeal costs so much that the cheaper route may be to pay off the large amount of lawyers' fees with a modest employee settlement payment. The lawyers keep between one-third and two-fifths of the money paid to employees as compensation for the illness.

7. The news media are very important participants in dealing with sick-building allegations. Rental and lease choices are made on perceptions; the job of the media is to establish and influence perceptions, so the quality of work that the owner does with news reporters can directly affect the value of the building in the future. Our chapter on communication with the news media is a useful guide to handling these responses.

8. Government tends to be reactive, not reflective, to worker allegations about health problems. A government agency would not decide to react harshly against a building based on complaints if the owner had adopted a prudent approach to communicating with the agency and cooperating factually with the technical shortcomings of the particular public agency. A response should always consider that owners and consultants have facts that government needs. The consultants brought in by the owner probably have much more expertise than the local or state health official, who might never have worked the diagnostic and causation aspects of such a claimed outbreak of illness in this official's past work.

9. Human resource professionals should be the primary interface between the complainant and the consultant and technical crew. A team approach is essential. The answers to technical questions require careful planning; once the answers are derived from data, the right way to communicate the answers is for the personnel manager or other experienced human resources person to put the answers into the kind of communication that the listeners will readily understand.

10. Showing respect for the dignity and intelligence of the individual employee will go a long way to avoiding legal conflicts. It is often wiser to make a minor adjustment than to fight a prolonged battle. The individual who senses concern and caring attention to his or her complaint often will respond with appreciation; the person who senses stonewalling apathy often will respond with hostile anger and demands for cash payment. Because this field has so many uncertainties of causation and effect, it is in the best interests of all that the first-level complaint should be met with friendly and prompt responses.

The reader is cautioned that medical and legal advice should be sought from appropriate professional advisers. This book offers general advice, to be applied with care in specific situations. More specific and sometimes different considerations will apply to any particular case that might arise. You should read this text, heed its preventive and planning advice, and then work with well-qualified consultants to tailor the response to any particular case.

About the Authors

James T. O'Reilly is Adjunct Professor of Law, University of Cincinnati.
Philip Hagan is Director of Safety and Environmental Management, Georgetown University, Washington, D.C.
Ronald Gots, M.D., is Principal, International Center for Toxicology and Medicine, Rockville, Maryland.
Alan Hedge, Ph.D., is a Professor in the College of Human Ecology, Cornell University, Ithaca, New York.

The authors express special thanks to contributor Tamara Lee Ricciardone of the Electric Insurance Company, Beverly, Massachusetts, formerly of Smith & Duggan, Boston, for her work on the Brigham & Women's Hospital case described in Chapter 15 and to contributors Robert Graham and Cynthia Drew of Jenner & Block, Chicago, for Chapter 17. Philip Hagan would like to thank Richard Pifer and Tom Edgerton for their wholehearted support of Safety and Environmental Management at Georgetown University, Chung-an Lin for performing many IAQ investigations that helped develop the basis for the practical experience that contributed to the idea of this book, and Cynni Nazeri and Carol Horn for the time they contributed in the research arena and other support areas.

Chapter 1

Does This Building Have an Indoor Environmental Quality Problem?

PHILIP HAGAN, MPH, CIH, CHMM

INTRODUCTION

This chapter addresses the technical evaluation of complaints of unsatisfactory indoor environmental quality (IEQ) and the creation of an effective IEQ plan for a building. We examine the likely causes for complaints and suggest methods of control. Appendix D provides lists of groups that may be of further assistance.

WHAT IS AN INDOOR ENVIRONMENTAL QUALITY PROBLEM?

The energy conservation measures enacted in the 1970s resulted in tighter buildings, lower requirements for ventilation, and subsequent increases in occupant complaints about poor health due to conditions of unsatisfactory indoor air quality (IAQ). Initially, poor ventilation was considered the primary culprit. The connection seemed straightforward: since IAQ complaints increased after the ventilation rate requirements fell, a reasonable assumption would be that increases in the ventilation rate should help address concerns about unhealthy indoor air. One widely reported study, from the National Institute for Occupational Safety and Health (NIOSH), asserted that most IAQ problems (54 percent) were caused by inadequate ventilation. However, in the past few years, analysis of data from numerous IAQ investigations has resulted in heightened awareness of multiple causes (improper building maintenance, microbial and chemical contamination, and excessive levels of volatile organic compounds [VOCs] from a variety of sources) that have added new dimensions to the investigation of indoor environmental problems. When complaints are received from building occupants, an investigation of likely causes should cover the entire indoor environment. While air quality is the factor most often implicated in IEQ investigations, noise, light, water quality, and other components of the indoor environment should be evaluated to determine if they are the basis of the complaint.

Investigations of likely causes and solutions of IEQ complaints can be as simple as identifying and cleaning up spilled cleaning material or as complex as a temperature inversion concentrating pollutants discharged from a neighboring building and funneling them into your building's air intake, where they are then distributed throughout the occupied spaces.

When building-related diseases are diagnosed, they can usually be identified and remedied. But many of the symptoms people complain about are diffuse and no specific etiological cause can be identified; these complaints can be collectively described as sick-building complaints. Building managers receive complaints, either specific or general, of unsatisfactory indoor environmental quality. The specific complaint tends to have a particular acute event as its cause, such as a broken water pipe that floods a carpeted area. The problem is readily identifiable and is able to be resolved promptly. General complaints tend to be more difficult to identify and quantify, because their causes are unclear.

Successful resolution of a complaint involving IEQ is not simply a matter of the building's maintenance staff increasing the ventilation to the complaint area. Building managers must observe all the stressing influences that could affect occupants in the particular area where the complaints occur. Too much noise, inadequate lighting, ergonomic problems, and job-related stress may be contributing to a problem that is vaguely, but incorrectly, attributed to air problems. A later chapter deals with the psychological issues that contribute to these less-specific complaints.

DEFINITIONS

The first term that should be defined when discussing indoor environmental quality is *health*. The World Health Organization (WHO) defines health as a "state of complete physical, mental and social well being, and not just the absence of disease or infirmity" (WHO, 1946). The goal of good IEQ should be to preserve the physical health and to promote the complete well-being of building occupants, which includes providing a reasonable degree of comfort. Concerns about comfort sometimes precede and underlie reports of physical illness. Because the indoor environment affects the total health of occupants, addressing all illness and comfort concerns at one time seems an appropriate management choice.

There *are* documented instances of illness caused by indoor conditions. The term *building-related illness* (BRI) is used to describe an illness caused by exposure to environmental agents in the building's air that cause identifiable symptoms of diagnosable illness, such as infections or certain allergies (Environmental Protection Agency [EPA], 1991), or specific medical conditions of known etiology that can often be documented by physical signs and laboratory findings (*Federal Register,* vol. 56, no. 183). BRIs are usually characterized by clinical signs, a low rate of occurrence to occupants, and prolonged recovery times after the individual leaves the building. Clinical signs may include fever, infection, tissue deterioration, blood abnormalities, and so on.

Building-related illness is addressed by removing the source of the cause of the illness, not simply improved ventilation. Examples of BRI include Legionnaires' disease from bacteria; toxic effects from exposure to asbestos, radon, carbon monoxide, and mycotoxins; and "humidifier fever" from fungi and bacteria.

Sick building syndrome (SBS) differs from BRI because its causes are not easily recognized and the method of mitigation is elusive. SBS is typified by complaints of acute discomfort, such as fatigue, congestion, headache, and eye irritation, by a significant percentage of the occupants of a building, with immediate relief of symptoms when the occupants leave that building (Walkinshaw, 1988). Respiratory complaints, fatigue, and irritation are typical, but many different symptoms have been associated with SBS. SBS describes cases in which the symptoms appear linked to the amount of time individuals spent in the building but for which no specific cause or diagnosable illness can be identified (EPA, 1991). It also refers to a class of complaints or symptoms generally not traceable to a specific substance but sometimes attributable to exposure to a combination of substances or individual susceptibility to low concentrations of contaminants (*Federal Register,* vol. 56, no. 183).

Causes of SBS can be very diverse, as discussed in the later chapters on medical and psychological aspects of the affects of the indoor environment. Problems can be caused by environmental stressors (overheating, noise, poor lighting), job-related psychosocial stressors (crowding, labor and management disagreements), ergonomic stressors, or unknown other factors (EPA, 1991).

According to WHO, SBS includes situations where a large percentage of the building population reports irritation of eyes, nose, throat, and skin; abnormal odor and taste; and nonspecific hypersensitivity or general health problems (WHO, 1983, 1984). The WHO group did not attribute to SBS those symptoms characterized by unusual individual sensitivity to indoor pollutants.

MANAGING IEQ

Indoor environmental quality issues are a potential problem for any building owner or manager. To help minimize the potential impact on building activities, the owner or manager should have a building IEQ management plan that addresses the management of the indoor environment. This plan should inform employees about indoor environmental issues and how any complaints will be handled. All IEQ-related complaints must be taken seriously, and a timely response is essential. If there is a real problem concerning building-related illness, a fast response and prompt mitigation will minimize any health concern before it becomes even more serious.

A successful IEQ plan should include a designated central contact person along with designated alternates. Resolving complaints would be very difficult with no central point of contact to collect and evaluate information. The building owner should recognize three benefits of central response: first, having a defined point of contact and response reduces alarm and hostility among those with complaints. Second, it assures that complaints are taken seriously by a person who has particular awareness of IEQ problems and how to resolve them. Third, it avoids rumor, panic, and distrust that can result from employees' unfounded speculation.

The central contact person would be assigned the title of *on-site IEQ coordinator*. All communications to the affected employees should be issued through this coordinator, not ad hoc from many different workers or managers.

The IEQ coordinator, who may be a manager or a consultant for the building owner, should be a credible person with good communication skills. The person should be directly responsible for understanding and implementing the IEQ plan. When IEQ complaints arrive, this person should direct the response effort and should communicate to all parties the status of follow-up and remedial measures. A prudent building owner will empower the IEQ coordinator to make decisions regarding the necessity of bringing in services for investigation and to initiate immediate remediation measures. Direct responsibility for handling complaints and addressing customer needs should be part of the IEQ coordinator's functions.

Most well-handled IEQ complaints end after the initial contact, once the IEQ coordinator has gathered the basic investigative information and performed the follow-up steps required for mitigation. A systematic, consistent approach asks basic questions of who, what, where, and when in response to employees' concerns.

The IEQ coordinator must be attuned to both the personal contact *and* the documentation requirements of the job. Standardized forms can assist in the speed and simplicity of the complaint-handling process. A standardized form should be clear and should be acceptable to legal counsel, because it will be crucial evidence in compensation and/or civil liability cases. This book will give suggestions that reflect acceptable industry standards for IEQ investigations. The building owner should be aware that in addition to spearheading the IEQ investigation, the IEQ coordinator is the first line of management "defense" against the more extreme compensation, litigation, and class-action liabilities that are discussed in later chapters of this text.

Written IEQ Guidelines

An internal IEQ building-management plan should establish written guidelines. These will be the procedures for the IEQ coordinator, building manager, facility engineers, maintenance and housekeeping personnel, and related employees to follow as the IEQ complaint is investigated. Forms from *Building Air Quality: A Guide for Building Owners and Facility Managers,* an EPA/NIOSH publication, are in the public domain and can be modified to fit your particular circumstance.

The introductory portion of the guidelines should define common terms and explain the common signs, symptoms, and causes of IEQ complaints. Individual responsibilities for each staff member involved in the IEQ investigation process should be explained.

The IEQ plan should detail the building's occupant densities and the minimum outside air ventilation rates. It should lay out the scenarios for possible complaints. It should describe the activities in the building that would be likely to affect IEQ, such as housekeeping and maintenance activities, removal of old furnishings, painting, roofing, ventilation changes, reconstruction, and so on, and should summarize the potential impact each factor may have on IEQ as well as how the adverse effect will be planned for and minimized, for example, how dust will be controlled during demolition and renovation projects. These guidelines should be used to proactively avoid IEQ problems. It is always more cost-effective and better for customers if problems resulting from poor indoor environmental quality do not occur.

A short, simple IEQ plan, made available to occupants and building workers, should cover the following topics:

- Proactive, preventive measures to protect IEQ
- How communications will be handled
- Responsibilities of the building owner
- Responsibilities of the building-management staff
- Responsibilities of building occupants and tenants
- Education of occupants about IEQ
- Description of building systems and their operating parameters
- Operating and maintenance activities that impact IEQ
- Heating, venting, and air-conditioning (HVAC) schedules that relate to building use

- Potential outdoor pollutant sources
- Potential indoor pollutant sources
- Potential pathways for pollutant entry

Written Operations and Maintenance Guidelines

The IEQ plan should be supplemented by a written operations and maintenance guidance document that includes building operating parameters, component upkeep through preventive maintenance, and training requirements. The emphasis of this document should be on preventing problems.

A proactive, not reactive, maintenance program is needed, because typically, preventive maintenance is more cost-effective than replacement cost of equipment that is not well maintained. Prevention and preemption of events leading to complaints saves time, limits costs, and avoids potential liability. A bearing that wears out in a ventilation fan due to lack of scheduled lubrication results in failure of the ventilation system, complaints from odors associated with the destruction of the bearing and the high cost of replacing the affected components. The benefits to the indoor environment will be maximized when the preventive maintenance helps equipment to operate closer to its original design parameters.

The written maintenance program is necessary because the levels of building-system maintenance can vary dramatically due to staff turnover, limited budgets, and complicated building systems. When the maintenance of complicated building systems is assigned to an untrained person, lack of adequate guidance and documentation can result in significant problems. Written maintenance programs with schedules for detailed preventive maintenance procedures can help avoid these problems.

Communications

The single most important response action to an IEQ complaint is to maintain open lines of ongoing communication. The process of response to and resolution of an IEQ complaint should never be a mystery to the building occupant. Without open lines of communication, distrust and inefficient waste of time and resources can be costly for the building manager or owner. Resolving complaints to everyone's satisfaction would be impossible without effective communications.

Ideally, lines of communication between building occupants and building manager will have been established before the IEQ complaint arises. Education about the building's systems and their function, how to deal with concerns, and definitions of good IEQ should be part of the information that aids in the resolution of future complaints and alleviates some of the fear of the unknown that may arise. A little money spent early on this type of communication can prevent dissatisfaction later, when untangling a misunderstanding is much more difficult.

Resolving an IEQ concern requires listening to the complainant, caring about that person's feelings, informing him or her about how the complaint is being addressed, documenting the steps taken, and following through on promises. The communication to occupants will contain information from three areas: risk assessment during the investigation; risk management, the identified response measures for dealing with the assessed risk; and risk communication, how and when information is presented to interested parties.

Occupants' perceptions and reality often differ in IEQ cases. Communication will deal with correcting misperceptions, from the first complaint through to the end of the process. As early as possible in the investigation, the scope of the response and an anticipated endpoint resolving the complaint should be identified and documented.

The communication of information to occupants by the IEQ coordinator should include the following:

1. Identification of the complaint area within the building
2. Progress of the investigation
   - Any factors that have been evaluated and found to be contributing factors
   - Any factors that have been evaluated and found *not* to be contributing factors
3. How long the investigation might take
4. Actions taken to improve the current situation
5. Status of future response actions

The occupant should be told of the mitigation strategy and also advised of the estimated time for the contaminants to disappear, since IEQ remediation usually does not have instant results. Thanking occupants for their patience and cooperation with investigators is also prudent.

To forestall predictable complaints, advance written notice of scheduled renovation projects and major HVAC maintenance should be given to occupants in the affected areas of the building. Notifying occupants of procedures such as construction, removal of walls or ceilings, floor stripping, and painting as a courtesy will preempt complaints.

An essential part of the communication plan is defining the mechanism for communication during an emergency. Plans for normal preventive maintenance and repairs are routine; plans for emergency situations should be flexible and should focus on the communication to occupants.

The plan will list "what if" scenarios and responses. For example, an emergency may be a flood in a carpeted work area. Cleaning and drying the carpet within 24 hours would be the response outlined in the plan. The benefits of this plan are twofold: Quick response to the flood prevents adverse effects on the building environment by preventing mold growth and odor problems, and occupants feel that responsive management addressed their concerns in a timely manner.

Building Partnerships on Risk Issues

One way to disarm hostility or lack of trust is to encourage a partnership between building managers and occupants to solve concerns. Building IEQ coordinators want to be seen as open, empathetic, competent, and committed to workable solutions. Reporting diaries and complaint forms can be used to inform building IEQ coordinators about the time of odors, discomfort complaints, and so on. Follow-up actions in response to the complaints are then also logged in the diary. This method helps occupants feel they are a part of the resolution process while it documents management response to concerns.

The building owner should give careful consideration to the selection of a public spokesperson and a person to communicate with employees. The individual(s) relaying information to the various parties should enjoy a certain level of trust and credibility. Employees' acceptance of recommended solutions resulting from an investigation will be affected by this level of trust and credibility. A 1991 EPA study referenced by Dr. Vincent Covello, School of Public Health, Columbia University, in the text *Protecting Public Health and the Environment* categorized the trust and credibility of sources of environmental information as shown in Table 1–1.

What constitutes "risk" varies according to different people's perceptions. Employees' trust in the person who relays information is enhanced when differences in risk perception are clearly addressed. The recommended solution(s) resulting from the investigation should be explained in a way that enhances credibility: *we found a certain cause, made a certain correction, monitored a certain parameter, analyzed certain samples, and our solution will be XYZ.* The likelihood that *XYZ* will be accepted depends on the credibility established through the education and communication process. Remember that industry sources tend to fall in the bottom third of public acceptance for credibility of environmental information, whereas physicians, professors, and nonprofit voluntary health organizations are in the top third. The media and environmental groups are in the middle (Covello, 1991).

Complaints of Thermal Discomfort

The IEQ issues that deal with thermal comfort are readily addressed. Associated with heat or humidity, they are usually easily identified and resolved. Building

Table 1–1. 1991 Sample Survey of the U.S. Population on Trust and Credibility of Sources of Environmental Information*

Top Third

Physicians and other health or safety professionals

Professors

Nonprofit voluntary health organizations

Middle Third

Media

Environmental Groups

Bottom Third

Industry Officials

Federal government officials

Environmental consultants from for-profit firms

Changes from previous years

Environmental groups: 10%–15% loss of credibility

Media: 5%–10% gain of credibility

Government and industry: 10% loss of credibility

*Adapted from *Protecting Public Health and the Environment.*

systems are designed to adjust the interior environment to acceptable levels of occupant comfort, taking into account uses, occupancy levels, and external weather conditions. Chronic complaints of thermal discomfort (too hot, cold, dry, or humid) could result from occupants attributing their own unrelated health symptoms to components of the indoor environment.

Complaints due to thermal and humidity conditions can be minimized by maintaining thermal environmental conditions in the ranges recommended by the American Society of Heating, Refrigerating, and Air Conditioning Engineers (ASHRAE) Standard 55, Thermal Environmental Conditions for Human Occupancy. In the winter, temperatures should be maintained between 68° and 75°F; in the summer, between 73° and 79°F. Humidity is the amount of water vapor contained in the air. Measurements are taken of relative humidities (percentage of moisture in the air relative to the amount the air could hold if saturated at the same temperature (EPA, 1991). Most people consider relative humidities between 30 percent and 70 percent comfortable. However, relative humidity above 50 percent can result in conditions conducive to microbial growth.

TRIGGERS FOR IEQ INVESTIGATIONS

Typical complaints attributed to poor IEQ include allergies, infections, eye irritation, fatigue, headaches, general malaise, and skin problems. Usually, the building manager or industrial hygienist lacks the tools, training, and knowledge to diagnose the basis for an individual's health concerns. Unless the source of an acute health effect has been identified and the response action has alleviated the medical problem, the medical complaint should be addressed by a physician, preferably one with training and experience in occupational medicine.

Information from the medical evaluation should be used to help identify the potential sources of the problem. In some cases, if support information from a physical examination is not available, resolving IEQ complaints based on occupants' health symptoms alone could be difficult. For example, if an individual's symptoms are related to a medical condition, not the indoor environment, any building-related response options would obviously fail to alleviate those symptoms.

Odor: Problem or Nuisance

Odors have long been a subject of much debate and discussion. In a 1916 study, more than 400 different scents were tested on human participants, and the results formed a psychological attempt at classification. On the basis of apparent perceived similarities, it was concluded that there were six main odor qualities: fruity, flowery, resinous, spicy, foul, and burned. To date, no unique chemical or physical property that can be said to elicit the experience of odor has been defined. Only seven of the chemical elements are odorous: fluorine, chlorine, bromine, iodine, oxygen (as ozone), phosphorus, and arsenic. Most odorous substances are organic (carbon-containing) compounds in which both the arrangement of atoms within the molecule as well as the particular chemical groups that comprise the molecule influence odor.

Unusual odors breed IEQ complaints. Odors are difficult to quantify and at times can be difficult to identify. Although odors are caused by airborne VOCs, the correlation between indoor VOC concentrations and indoor environments is poor (ASHRAE, 1993). To be odorous, a substance must be sufficiently volatile to give off molecules and be carried into the nostrils by air currents. The solubility of the substance is probably a factor, since chemicals that are soluble in water or fat tend to be strong odorants, although many of them are not. Temperature and humidity influence the strength of an odor by affecting its volatility and hence the emission of odorous particles from the source and the resulting dissipation of the odor.

Odor can signal a serious problem or merely indicate a nuisance. When odor complaints are received, education and prompt response are the best defense. A quick response should identify the odor and determine the level of risk so that appropriate measures can be initiated. Prompt response to an odor facilitates complaint resolution. The response should evaluate the potential sources inside and outside the building, check the ventilation system, and examine the activities of building occupants. Education about the actual source found, as well as its remedial measures, will reduce anxiety about the strange odor. If the odor information is not communicated, then the occupants may worry that the odor source is the cause of unrelated individual symptoms.

In recent years, researchers have developed sensory analysis techniques and definitions in their attempt to quantify odor sources (Fanger, 1988). The quantification of odors uses the term *olf,* defined as the emission rate of air pollutants emanating from one standard person. A unit of pollution is defined as the pollution generated by 1 olf ventilated by 1 liter per second of supplied fresh air. To evaluate perceived air pollution, a panel of more than 50 trained judges evaluates a test environment under three different operating scenarios. Results are analyzed by evaluating responses to a questionnaire. Because the method of assessment is subjective, not objective, this measuring tool would appear to be more of a research instrument and less likely to be selected for a real-world IEQ investigatory technique.

The World Health Organization describes a threshold value for nuisance odors and the concentration at which less than 5 percent of the population would experience annoyance less than 2 percent of the time (WHO, 1987). The American Society of Heating, Refrigeration and Air Conditioning Engineers (ASHRAE) sets a criterion of acceptable air quality if 80 percent of an untrained panel of 20 observers considers the air to be "not objectionable" upon initial entry into an area. Initial entry is specified because after a short exposure time, the body's olfactory nerve (sense of smell) experiences fatigue from the odor's effects (ASHRAE, 1989). Another criterion used to measure odor is based on carbon dioxide, a by-product of breathing. ASHRAE recommends an indoor air limit of 1,000 parts per million (ppm) of carbon dioxide to satisfy its odor criteria (ASHRAE, 1989).

Education of building occupants is important, in particular the relationship between odors and the toxic effects of chemicals. Some substances are toxic with no odor, such as carbon monoxide gas. Some odors have no toxic effect. Other odors, such as methylmercaptan in natural gas, provide excellent warning of the presence of toxic chemicals.

Although masking can reduce an odor's impact, it is less effective than chemical conversion or physical collection (removal) as a basis for control. Some odorants also may be removed by passing air through activated charcoal. If this is used, then care should be taken to ensure the capacity of the filter is not exceeded, which could result in a rerelease of contaminants.

The prudent building manager will include in the IEQ management plan a catalog of substances used in the building that could potentially result in odor complaints. In the event of such a complaint, this list can aid in more rapid identification of a probable source.

RESPONSES BY THE IEQ TEAM

Although most indoor air quality investigation protocols use similar strategies, the lack of specific regulatory requirements and the intricate nature of IAQ problems can result in different approaches for each investigation. Dr. Phillip Williams, associate professor in the environmental health science program at the University of Georgia, recently published results of a study classifying IAQ investigation protocols by their use of four general assessment methodologies: solution-oriented, building diagnostics–oriented, industrial hygiene–oriented, and epidemiological (Williams and Greene, 1996).

The most common strategy identified was a solution-oriented approach. This investigation protocol is based on the systematic exclusion of a narrowing range of possibilities and involves a preliminary assessment, a walk-through investigation, measurements of applicable parameters, and a report summarizing findings and recommendations. The U.S. EPA publication *Building Air Quality: A Guide for Building Owners and Facility Managers* uses a solution-oriented approach in describing methodologies for investigating an IAQ complaint. Medical diagnosis and input can be used to validate specialized sampling when measurements are conducted beyond basic environmental parameters.

The use of building diagnostics involves an initial consultation stage very similar to the preliminary assessment and walk-through of the solution-oriented approach. This is followed by qualitative diagnostics that establish performance standards (e.g., for the indoor environment, HVAC system, etc.), characterize complaints, and perform sampling for specific pollutants. The last stage, quantitative diagnostics, involves sampling sites based on potential impact to the occupant, implementing a quality assurance and control program for monitoring, gathering data through the use of a questionnaire, and analyzing and interpreting data, followed by recommendations.

Industrial hygiene protocols are based on recommendations from two industrial hygiene organizations, the American Industrial Hygiene Association (AIHA) and the American Conference of Governmental Industrial Hygienists (ACGIH). Both sets of recommendations are very similar in nature to the solution-oriented approach. However, the ACGIH protocol specifically addresses complaints and sampling in relation to bioaerosols.

In epidemiological protocols, patterns of the occurrence of disease as related to time, place, and person are used to formulate and test hypotheses.

Regardless of the type of protocol used in pursuit of a solution to an IAQ problem, a multidisciplined team approach probably has the best chance of providing successful resolution.

A generic synopsis of the various IAQ protocols would include the following steps:

1. *Initial contact.* Initiate indoor environmental evaluations in response to complaints from affected parties or supervisors. Evaluations of affected persons by medical personnel can result in requests for evaluation of the indoor environment.
2. *Preliminary assessment and problem review.* Collect background information to develop a starting point for the investigation: query occupants (employees in both complaint and noncomplaint areas, management, building engineers, housekeeping). If possible, develop a profile for symptoms (timing, type, location).
3. *Visual evaluation and walk-through.* Evaluate status of normal operational functions (maintenance, housekeeping, building use scenarios), HVAC operation, nonroutine operations (renovation, maintenance, housekeeping), potential pollutant sources (inside and out).
4. *Simple industrial hygiene measurements.* Carbon dioxide level, temperature, and relative humidity must all be measured.
5. *Specialized testing.* Perform tests to identify VOCs, microbial contamination, and specific chemical contaminants. This type of testing should be performed only if there is a valid reason for doing so. Medical evaluation and elimination of a suspect contaminant source are two reasons for conducting specialized sampling. Sampling goals and objectives along with appropriate responses should be identified beforehand.

Employees or consultants on the IEQ response team should represent all the necessary disciplines, including environmental health, mechanical engineering, building design/architecture, facility maintenance, operations, engineering, contract service providers, public relations, personnel, and housekeeping. The IEQ management plan should identify which persons will respond and what levels of response will be indicated during each phase of the response effort.

Unfortunately, because of poor planning or poor management awareness of needs, often a full IEQ response team is assembled only when a critical emergency already exists at the building—making a well-reasoned, amicable resolution of complaints much more difficult. At that stage, psychologists should be included on the team, because emotions drive the conflict and perceptions have become confused with realities.

The building owner's cost-reduction goals are best met by preplanned availability of contractors. For example, it can be arranged that experienced medical and legal service providers will be available on call at identified prices. The availability of expertise from local universities can be explored. Public officials can be located who could provide informal recommendations and information. In-house resources usually need to be augmented with consultant service providers during the short time necessary to identify and resolve complex IEQ scenarios.

Practice sessions, familiarization visits, and discussions between in-house and consultant members of an IEQ response team can help prepare team members for solving actual complaints when they occur. Preplanning particularly helps a team deal with inaccurate perceptions before they become reality and before emotions dominate the thinking process of individual complainants.

Responses through In-House Staff

The benefit of an in-house IEQ team is that intimate knowledge of the facility's internal workings can hasten the identification and remediation of problems. The in-house staff member has knowledge of history, past uses, and present occupants that would take consultants considerable time to piece together. Most simple IEQ complaints can be resolved by properly trained in-house staff. A disadvantage is that quality of experience grows from dealing with multiple problems in multiple buildings. Consultants are more likely to have this breadth of experience. Another disadvantage is that occupants might perceive outside consultants as more credible than familiar building staff members. During the gathering of data, the in-house team may not have sufficient knowledge to analyze the collected data and speak with assurance about causes and effects. However, a fully staffed, experienced team of consultants dedicated solely to handling IEQ complaints is a luxury that many building owners cannot afford.

Responses through Consultants

Consultants who specialize in IEQ have usually performed enough inquiries to have specialized knowledge that can help them when analyzing the data collected during the investigation. They profit from experience in several ways.

They can shorten the time needed to conduct an investigation by applying logic and investigative techniques that were successful in other buildings.

In general, a consultant who has had a wide range of IEQ experiences in other settings can search out clues that others might miss. Investigating other IEQ problems in other sites has taught this person how to diagnose situations more rapidly than someone who is a relative novice. A good consultant will bring to the table a wide range of experiences that can minimize time spent looking for ghosts. Additionally, a consultant who is on the payroll will have dedicated time to pursue the project.

Disadvantages of relying on consultants for IEQ assistance include higher costs of investigations. Cost can be controlled to some degree if the IEQ problem is first handled by in-house personnel, and outside consultants are used only for bigger and less-understood IEQ complaints.

A second disadvantage is that consultants sometimes protect themselves by recommending the most elaborate remedial measure. This can be problematic if the cost of that option exceeds the building manager's ability to pay. A way to minimize this disadvantage is to ask the consultant to supply alternative options for remediation.

A third concern is the communication of results. Some consultants have difficulty giving plain-language advice. The building manager and occupants need to understand what is being offered. Checking references can ensure that the consultant is capable in this area.

A fourth issue is specificity of skills. Consultants tend to have tunnel vision dictated by their particular area of expertise. When selecting consultants to assist in your IEQ management program, it is important to verify that their skills contribute positively. If an industrial-hygiene consulting firm provides an industrial hygienist whose primary experience is in monitoring asbestos-abatement projects or who has recently graduated from college and is lacking in work experience, that individual will probably not be able to make a significant contribution to the overall IEQ team effort. Any consultant you hire should have experience dealing with a wide range of IEQ investigations. By the same token, when problems are identified that require a specialist, do not hesitate to hire the necessary expertise.

When consultants are used, contracts should detail expectations and the scope of work. Wants and needs of both parties should be spelled out before any work is done. Consultants should be asked to provide several solutions to an identified problem whenever possible. Insurance coverages, references, and previous experience in the IEQ field should be verified. Time frames for work and written reports should be laid out before work commences.

Carefully evaluate the IEQ consultant applicant's references. The following questions can provide good background information when checking references:

- Did the consultant help identify the problem and offer reasonable solutions?
- Were both management and the occupants happy with the communication process throughout?
- Were the terms the consultant used to describe the investigative process and results understandable and presented in a timely fashion?
- Did the consultant perform tasks in accordance with contract commitments?

Many local and state health departments can provide expertise that will be helpful in investigating IEQ problems, but the owner must realize that the news media's close relationship with local officials means there will be broad public awareness once the interactions at the local level begin; this is not so much a problem when federal agencies become involved. Appendix G provides information on where to obtain an EPA and Public Health Foundation publication, *Directory of State Indoor Air Contacts,* that can be used as a resource for state contacts during IEQ investigations.

BUILDING CODES

Indoor environmental quality issues interrelate with building-design issues, and the standards of design and operation are important for the building operator to understand. National building codes provide a series of model regulatory construction codes that are designed to be adopted by state or local jurisdictions. The codes are published in sections covering different aspects of building construction (plumbing, fire prevention, mechanical, etc.). The three primary code organizations are Building Officials and Code Administrators International (BOCA), the International Conference of Building Officials (ICBO), and the Southern Building Code Congress International (SBCCI). The stated purpose of the 1991 ICBO *Uniform Building Code,* "to offer minimum standards to safeguard life or limb, health, property and public welfare by regulating and controlling the design, construction, quality of materials, use and occupancy," shows that the general intent of the codes is to provide building-design criteria that will protect the health, safety, and welfare of the building occupants, incorporating IEQ.

For instance, all three major building-design codes in the United States mandate minimum outside air ventilation rates in mechanically ventilated buildings. ASHRAE Standard 62–1989 was adopted by the SBCCI in 1992 and was the basis for ventilation requirements in the 1993 versions of the *Uniform Building Code* and the *Mechanical Code of Building Officials and Code Administrators*

(ASHRAE Standard 62–1989 is discussed later in this chapter). However, unless a local jurisdiction has specified that the codes must be used for building operations, the codes are generally just building-design guidelines. Check the age of the building to determine what codes are applicable. If major renovation has occurred, some codes require retrofitting to meet applicable codes for the newer work.

FEDERAL GUIDELINES

Although several federal agencies conduct activities that impact IEQ, no single federal regulatory body has specific overall responsibility for the indoor environment. A later chapter addresses inspection and regulation issues in detail. Specialized building projects also have specialized jurisdiction that is beyond the scope of this text. For example, federal buildings are the responsibility of the General Services Administration, and federally funded public housing construction must comply with standards set by the Department of Housing and Urban Development (HUD). If your building is in either of these categories, it would be prudent to check with the appropriate federal agency's regional office for current IEQ information.

EPA, OSHA, and NIOSH

EPA, the Occupational Safety and Health Administration (OSHA), and NIOSH are the three major agencies whose activities can impact IEQ. EPA and OSHA are the two main U.S. federal agencies whose primary responsibility is to regulate environmental and personal exposure to hazardous materials. As a part of the Public Health Service, NIOSH serves as a research and support function concerning IEQ for the other agencies that interact with building owners.

The federal EPA conducts a nonregulatory indoor air quality program emphasizing research, technical guidance, and training. However, the EPA also issues regulations and directs other activities that affect indoor environmental quality under the laws for drinking water (Safe Drinking Water Act [SDWA]), pesticides (Federal Insecticide, Fungicide, and Rodenticide Act [FIFRA]), toxic substances (Toxic Substance Control Act [TSCA]), and outdoor air (Clean Air Act [CAA]). The CAA is not applicable to air in the indoor environment, but regulation of the outside air that is supplied to the building can impact the indoor environment.

Title IV of the Superfund Amendments and Reauthorization Act (SARA) was used to establish the Radon Gas and Indoor Air Quality Research Act of 1986. This act directed the EPA to establish an indoor air quality research program, to coordinate with other public organizations, and to disseminate information on indoor air quality to the public (EPA, 1989). The EPA Indoor Air Division under the Office of Air and Radiation is responsible for implementing indoor air policy and program development, and the EPA Office of Research and Development is responsible for the law's several IAQ research provisions.

EPA has used SARA and other statutes to study specific pollutants that could affect IEQ: radon, asbestos, environmental tobacco smoke, formaldehyde, chlorinated solvents, and pesticides. Information on these and other studies concerning IAQ is available from EPA's Indoor Air Quality Information Clearinghouse, P.O. Box 37133, Washington DC 20013-7133 (toll-free phone number, 800-438-4318). Each EPA regional office can be contacted for information on indoor air, radon, and asbestos. Telephone numbers for EPA's regional offices are listed in Appendix C.

OSHA was created by the Occupational Safety and Health Act in 1970. Its requirements apply directly to private-sector employees and indirectly through state laws and federal executive orders to public-sector workers. OSHA promulgates safety and health standards, provides consultation to employers, conducts inspections, and enforces regulations to ensure that workers are provided with safe and healthful working conditions. Employers with fewer than 10 employees are basically exempt from OSHA inspections except in cases of complaints or accident investigations. Phone numbers for OSHA regional offices are listed in Appendix C.

OSHA regulates worker exposure to hazardous materials with specific standards: 29 CFR 1910 (General Workplace) and 29 CFR 1926 (Construction). Standard 29 CFR 1910 outlines permissible exposure limits (PELs) to hazardous chemicals. PELs are intended for use in evaluating exposure by healthy working adults in industrial settings, and their applicability to nonindustrial environments is a matter of much discussion. They may not adequately reflect the health realities of today's nonindustrial working population. Care should be taken in applying these standards to a nonindustrial building environment.

OSHA is authorized to conduct workplace inspections and can enter a workplace upon the owner's consent or with a search warrant. Employers may refuse entry if an inspector does not have a search warrant. Chapter 8 addresses inspection issues.

Employee complaints of poor IEQ are one of the mechanisms by which OSHA is triggered. At one time, OSHA responded to each complaint by instigating an on-site inspection. Recent budget shortages, however, have produced a variation on that approach. Often, when a complaint is received, instead of inspecting the site, OSHA's local office sends the employer a letter. The letter is in the form of a notice of alleged violation and asks the employer to address and respond to the complaint. The employer can then investigate and send a response to OSHA detailing the investigation and any corrective actions taken. If OSHA is not satisfied with the response, the agency may conduct an on-site inspection. Employees who contact OSHA with complaints are protected by law against employer reprisals such as termination or discipline.

Since there are no specific OSHA standards governing IEQ, OSHA inspectors would have to turn to the general duty clause. OSHA's general duty clause states the following:

Sec. 5(a) Each Employer—

Sec. 5(a)(1) shall furnish to each of his employees employment and a place of employment which are free from recognized hazards that are causing or are likely to cause death or serious physical harm to his employees;

Sec. 5(a)(2) shall comply with occupational safety and health standards promulgated under this Act.

Sec. 5(b) Each employee shall comply with occupational safety and health standards and all rules, regulations, and orders issued pursuant to this Act which are applicable to his own actions and conduct.

Citing a violation of the general duty clause requires the existence of certain elements. First, there must be a recognizable hazard; second, the hazard must be likely to cause death or serious physical harm; and third, OSHA must be able to show that the hazard can be abated by a feasible and useful method. If OSHA finds these elements, it proposes a penalty and the case goes before an independent judge and finally to a review commission that is independent of OSHA and that to date has not adopted a legal standard policy for IEQ. Historically, OSHA has assessed administrative fines under its general duty clause for companies that do not comply with air quality regulations. As of 1997, OSHA has never used the general duty clause to cite employers for poor indoor environmental quality.

NIOSH, a part of the Public Health Service, conducts research, supports educational research centers, conducts health hazard evaluations of the workplace, furnishes support documentation to OSHA during the standards-development process, and conducts training on various issues, including indoor air quality to promote safe and healthful workplaces. NIOSH will consider requests to conduct investigations when solicited by employees, employers, other federal agencies, and state and local agencies to identify and mitigate workplace problems. Requests for field investigations can be addressed to NIOSH, Hazard Evaluations and Technical Assistance Branch (R-9), 4676 Columbia Parkway, Cincinnati, OH 45226. Requests for information are handled at 800-35-NIOSH or 800-356-4674.

EPA and NIOSH have developed solution-oriented guidelines for building owners to follow when investigating an IEQ complaint: *Building Air Quality, A Guide for Building Owners and Facility Managers,* EPA/400/1-91/033, DHHS (NIOSH) Publication no. 91-114, December 1991.

Other federal agencies with indoor air responsibilities are listed in Appendix B.

STATE REGULATORY PROGRAMS

States have a role in responding to IEQ concerns. Some state health officials have greater expertise in dealing with IEQ than their federal counterparts. Concerns about radon and environmental tobacco smoke have resulted in legislation and health regulations on state and municipal levels over the past 10 years.

Every state has at least one agency that is responsible for IEQ issues. Most states have information and educational programs in place. Some have initiated programs that address specific parameters affecting IEQ. Parameters commonly covered by these programs include ventilation, radon, environmental tobacco smoke (ETS), lead, and asbestos. Some states, such as Florida, Illinois, and Oregon, have developed progressive programs that include criteria for acceptable IEQ and field services that investigate certain IEQ complaints.

California's Title 8, paragraph 5142 applies to indoor air quality in commercial workplaces and requires HVAC systems to be inspected annually, with any defects promptly corrected. It also requires that ventilation systems operate according to design parameters for supply air and that fans stay in operation during hours of occupancy.

Usually IEQ issues are addressed by more than one state agency. IEQ programs at city or county levels have gathered useful experience and can serve as a cost-efficient resource to employers. As mentioned earlier, however, a building owner who interacts with local officials must expect media attention that may be premature and inaccurate, so a discussion about the publicity of the IEQ inquiry should take place at the very start of interaction with local health officials (see Chapter 8). In some cases, IEQ issues have escalated into the public health arena. When this happens, state and county officials become involved, sometimes resulting in building evacuation and closure. Fire department hazardous material teams are often called in to respond to serious acute incidents, such as hazardous materials spills that impact the inside environment of a building, natural gas leaks, carbon monoxide poisonings, and other similar waves of nausea or fainting. Fire departments are not normally equipped or inclined to deal with routine IAQ situations, such as sick building syndrome or building-related diseases.

The EPA and Public Health Foundation publication *Directory of State Indoor Air Contacts* lists state agency contacts that can be a valuable resource during IEQ investigations. This directory can be obtained from EPA's Indoor Air Quality Information Clearinghouse (IAQ INFO), 800-438-4318 or 202-484-1307.

INDOOR ENVIRONMENTAL STANDARDS AND GUIDELINES

ASHRAE has used a multidisciplined team-review approach to develop and update several consensus standards (ASHRAE Standard 55, Thermal Environmental Conditions for Human Occupancy; ASHRAE Standard A90, Energy Management; ASHRAE Standard 62, Ventilation for Acceptable Air Indoor Air Quality; and ASHRAE Standard 111, Measurement, Testing, Adjusting and Balancing of Heating, Ventilation and Air-conditioning Systems) that can directly impact indoor environmental quality. ASHRAE normally uses a five-year cycle to review and revise existing standards to incorporate new knowledge and technology. Each succeeding revision includes the year of the revision in the title.

ASHRAE standards are voluntary standards. This means that requirements become enforceable only after a state or locality adopts the standard as its building code. Typically, building codes are used as guidelines in the design of buildings and their applicable systems. The codes are not enforceable during the later building operations. Some states are working on applying existing design codes and standards to building operations.

ASHRAE's first ventilation standard, ASHRAE Standard 62-73, provided a prescriptive approach to ventilation by specifying both minimum and recommended outdoor air-flow rates. This standard was referenced by ASHRAE's first energy standard, 90-75, and the revised 90A-1980 and is still referenced in many building codes.

In keeping with ASHRAE's use of a five-year review cycle, the 1981 revision of ASHRAE Standard 62 recommended different outdoor air-flow rates for spaces where smoking was allowed than for nonsmoking spaces. ASHRAE Standard 62–1981 also recommended procedures that allowed building engineers to implement innovative energy-management practices as long as concentrations of indoor air contaminants were maintained below certain limits. Due to some confusion in implementing the new standard, ASHRAE authorized an early review in 1983. The resulting revision, ASHRAE Standard 62–1989, recommended procedures that indicated a compromise between energy management and establishing acceptable indoor air quality.

ASHRAE Standard 62-1989 specified minimum outdoor ventilation rates ranging from 15 to 60 cubic feet per minute per person (cfm/person), which should result in acceptable indoor air quality for a substantial majority (greater than 80 percent) of the people exposed:

Location	Cubic feet per minute per person (fresh air)
Auditoriums	15
Cafeterias	20
Conference rooms	20

Kitchens	15
Libraries	15
Office spaces	20
Smoking lounges	60
Spectator sport arena	15

ASHRAE recently completed and released the newest revision of ASHRAE 62-1989: ASHRAE 62-1989R (Ventilation for Acceptable Indoor Air Quality) for public comment. This revision goes further than previous documents by recommending building operation and maintenance procedures to ensure that ventilation and building systems result in acceptable indoor air quality. To facilitate adoption of ASHRAE 62-1989R, code language was used to enable local building code authorities to incorporate the standard directly into their codes with a minimum of redrafting. The expressed purpose of ASHRAE Standard 62-1989 is to specify minimum ventilation rates and indoor air quality that will be acceptable to human occupants and "are intended to avoid adverse health effects."

The following are the general requirements of ASHRAE 62-1989R:

- Document design assumptions and calculations.
- Design for good air distribution.
- Locate outdoor air intakes properly and provide makeup air for indoor combustion.
- Design for easy air system cleaning and select appropriate filters.
- Provide local exhaust for stationary sources.
- Specify sloped drain pans with access for cleaning.
- Follow either the Ventilation Rate Procedure which stipulates minimum outdoor airflow rates for adequate dilution or the IAQ Procedure which specifies contaminant levels and subjective evaluation for acceptable indoor air quality.

The purpose of the next ASHRAE Standard, Standard 111–1988R, is to establish uniform procedures for measuring, testing, adjusting, balancing, evaluating, and documenting the performance of building HVAC systems in the field. The scope of this standard applies to building HVAC systems and their associated components. This includes (1) methods for determining thermodynamic conditions; (2) methods for determining electrical conditions; (3) methods for determining room air change rates, room pressurization, and potential cross-contamination of spaces; (4) procedures for measuring and adjusting outdoor ventilation rates to meet design specifications; and (5) methods for validating collected data. This standard also establishes minimum system requirements to ensure that systems can be field-tested and balanced. The field data collected and reported under this standard should be used by building designers, engineers, and users and by manufacturers and installers of HVAC systems. This information can be used to ensure optimal performance of the building's HVAC systems throughout their operational life.

Other ASHRAE standards that can impact IAQ are included in Appendix D. ASHRAE materials are available from the ASHRAE Publication Sales Department, 1791 Tullie Circle, NE, Atlanta, GA 30329; telephone 404-636-8400.

The ventilation directory supplied by the National Conference of States on Building Codes and Standards, Inc., 505 Huntmar Park Drive, Suite 210, Herndon, VA 22070; telephone 703-481-2020, is a useful resource that summarizes natural, mechanical, and exhaust ventilation requirements of the model building codes; unique state codes; and ASHRAE standards.

MITIGATION AND CONTROL STRATEGIES

After an indoor air quality problem is identified, four basic methods of control can be used for mitigation:

Source management
Ventilation
Exposure control
Education

When IEQ problems have been determined to be caused by IAQ, four primary control strategies can be used to prevent or control them: source management, ventilation, exposure control, and education.

Source management involves minimizing or eliminating exposures to chemicals (e.g., pesticides, cleaning agents, combustion by-products), allergens (e.g., dust mites, fungi, insect parts), or other potential sources of indoor environmental pollution. Source management uses substitution, enclosure, or encapsulation to control contaminants. Source management is generally the most effective control method. Substitution involves replacing a toxic substance with one that is less toxic (e.g., replacing a petroleum-based cleaning substance with a citrus-based one). Enclosure uses barriers to prevent or minimize the release of pollutants into the surrounding environment (e.g., walls enclosing a photocopier). Encapsulation involves covering a potential pollutant source to prevent the release of the pollutant into the surrounding area (e.g., sealing a wall that is covered with lead-containing paint).

Ventilation is used to maintain thermal comfort; to control contaminants by physical removal, dilution, or filtering; and to introduce fresh air into the building. Increased ventilation rates are often used for control when specific contaminants have not been identified. Ventilation can be used to remove pollutants before they build up to problem concentrations. Indoor air quality problems can be controlled through the use of ventilation to supply outside air to dilute indoor air pollutants. Ventilation systems can clean air by using filters to remove particulates or gaseous pollutants.

Exposure control removes building occupants from the pollutant exposure through the use of administrative management techniques (e.g., scheduling housekeeping operations during times when the building is unoccupied).

Education is a critical part of any IEQ management program. A later chapter deals with the psychological issues involved in communicating with occupants. When individuals are provided information about the potential sources and potential effects of pollutant sources in their control, they can use their knowledge of the workplace coupled with common sense to offset adverse effects. Building maintenance workers should be taught about source management and proper maintenance operations. Building workers can be made partners in the process by being taught to identify and report scenarios that could adversely affect IEQ. Building occupants should be informed of the elements of the IEQ management plan and whom to contact with any IEQ concerns (i.e., the designated IEQ coordinator). The elements of in-place IEQ programs should be communicated to building occupants. If building occupants are made aware of the existence of proactive preventive measures, their comfort level with building-management response actions to complaints will be higher.

Information on outdoor air quality parameters can be made available to occupants. The impact of high or low outdoor humidities, pollen counts, and smog days can have adverse effects on a person's health. High pollutant standard indexes (see "Written IEQ Guidelines" earlier in this chapter) can be communicated to building occupants on a regular basis.

Local incidences of cold and influenza (flu) outbreaks could be tracked and provided to building occupants. Cold and flu outbreaks are typically transmitted from person to person without regard to building systems. Many people are not aware of this and mistakenly attribute their cold and flu symptoms to a "sick building." The availability of flu shots prior to flu season should be publicized.

The Importance of Common Sense

Are the recommended solutions practical for this building? After the sources of an indoor air quality problem are identified and solutions for alleviating the cause(s) are recommended, both the short- and long-range elements of the solution should be evaluated to determine if they are practical. Redesigning a ventilation system could alleviate an indoor air quality problem, but so could increasing the system's operating hours or using preventive maintenance (e.g., cleaning or replacing filters, replacing worn parts) to ensure that the HVAC system is operating at the maximum level of performance. All parties involved in implementing the fix should be contacted for their input as to the practicality of implementation. If a new filter system is installed and maintenance workers are not able to access the equipment to inspect or change the filter, the "fix" will not be very effective.

Do the control measures correspond to the problem? Sometimes recommended control measures address only a portion of the problem. It is always important to select a control measure of a size and scope that fit the problem. Common sense should be applied liberally when identifying a workable solution to an IAQ problem. For example, if occupants are having problems with thermal comfort, replacing filters in the HVAC system with higher-efficiency filters will not help the situation.

Solutions to IAQ problems should make sense and should address the specific, identified problem(s). It is usually not a good idea to implement mitigation measures until problems have been identified. Otherwise, if the wrong mitigation measures are implemented and they do not solve the problem, those making the complaints tend to lose faith in the IEQ response team, resulting in bad feelings and mistrust toward the team as well as management personnel.

Is the solution permanent? An permanent solution is usually preferable to a temporary fix. If a weeklong renovation project results in complaints of thermal discomfort because part of the HVAC system has been temporarily shut down, upgrading fans in the HVAC system to increase the air supply to the area is probably not a reasonable fix. A possible fix would include providing temporary cooling to the area of concern. However, if the renovation project is going to last for an extended period (e.g., longer than a year), permanent upgrades to the HVAC system would be more appropriate.

Has the cost of implementation been fully evaluated? Both short-term and long-term costs should be examined. Short-term costs could include the cost of design, equipment, renovation, and labor for installation of any equipment. Long-term costs could include staff time, preventive maintenance, education, energy costs, and replacement parts. A solution that initially looks inexpensive could become prohibitively expensive in the long term. It is always a good idea to explore several possible solutions and their potential short- and long-term impacts.

At other times, the recommended control measures are cost-prohibitive. If changing maintenance or housekeeping procedures will solve the problem, there is no need to move to a new building or install a new ventilation system. On the other hand, if the problem is caused by a serious deficiency in the ventilation system, no amount of housekeeping changes will solve it. Recommendations of any kind should be examined very closely and implemented only if they will provide a reasonable solution.

Does the solution fit with the operation of the facility? Solutions that can be readily incorporated into a facility's existing operation generally gain easier acceptance and are more effective than exotic fixes requiring unfamiliar equipment and procedures. If the solution does incorporate changes in maintenance or housekeeping procedures, then education and training programs should be developed and maintained. Purchasing practices, checklists, and operating and maintenance schedules for facility operations may need to be modified.

NOTES

American Conference of Governmental Industrial Hygienists (ACGIH). 1989. *Air Sampling Instruments for Evaluation of Atmospheric Contaminants,* ACGIH, Cincinnati.

ACGIH. 1989. *Guidelines for Assessment of Bioaerosols in the Indoor Environment,* ACGIH, Cincinnati.

ACGIH. 1995–1996. *Threshold Limit Values and Biological Exposure Indices.* ACGIH, Cincinnati, Ohio.

American Industrial Hygiene Association Biohazards Committee (AIHA). 1993. *Biohazards Reference Manual.* AIHA, Fairfax, Va.

American Society of Heating, Refrigerating, and Air-Conditioning Engineers (ASHRAE). 1976. *Method of Testing Air-Cleaning Devices Used in General Ventilation for Removing Particulate Matter: ASHRAE Standard 52–76.* ASHRAE: New York.

ASHRAE. 1981. *Thermal Environmental Conditions for Human Occupancy: ASHRAE Standard 55-1981.* ASHRAE, Atlanta.

ASHRAE. 1989. Guideline 1-1989. *Guideline for the Commissioning of HVAC Systems.* ASHRAE, Atlanta.

ASHRAE. 1989. *Ventilation for Acceptable Indoor Air Quality.* ASHRAE, Atlanta.

ASHRAE. 1993. *ASHRAE Handbook: Fundamentals, Edition—1993.* ASHRAE, Atlanta.

ASHRAE. 1996. *ASHRAE Handbook: Heating, Ventilation, and Air Conditioning Systems and Equipment, Edition—1996.* ASHRAE, Atlanta.

Building Officials and Code Administrators (BOCA) International, Inc. 1989. *BOCA Basic/National Building Code.* BOCA International, Country Club Hills, Ill.

Council of American Building Officials (CABO). 1989. *CABO Building Code.* CABO, Falls Church, Va.

Environmental Protection Agency (EPA). 1985. *Asbestos in Buildings: Guidance for Service and Maintenance Personnel* (in English and Spanish). EPA 560/5-85-018, Washington, D.C.

EPA. 1985. *Guidance for Controlling Asbestos-Containing Materials in Buildings.* EPA 560/5-85-024, Washington, D.C.

EPA. 1987. U.S. Environmental Protection Agency, Office of Acid Deposition, Environmental Monitoring and Quality Assurance. Project Summary: The Total Exposure Assessment Methodology (TEAM) Study. EPA-600-S6-87-002, Washington, D.C.

EPA. 1989. *Indoor Air Facts* (No. 6). EPA Report to Congress on Indoor Air Quality. Washington, D.C.

EPA. 1989. *National Primary and Secondary Air Quality Standards.* Code of Federal Regulations, Title 40, Part 50.

EPA. 1989. *Report to Congress on Indoor Air Quality. Vol. II. Assessment and Control of Indoor Air Pollution.* EPA 400/1-89/001C. US EPA Office of Air and Radiation, Washington, D.C.

EPA. 1990. *Managing Asbestos in Place: A Building Owner's Guide to Operations and Maintenance Programs for Asbestos-Containing Materials.* Washington, D.C.

EPA. 1990. *Ventilation and Air Quality in Offices.* Indoor Air Quality Fact Sheet #3, Washington, D.C.

EPA. 1991. *Building Air Quality: A Guide for Building Owners and Facility Managers.* EPA/400/1-91/033, DHHS (NIOSH) Publication No. 91-114, Washington, D.C.

EPA. 1991. *Introduction to Indoor Air Quality: A Self-Paced Learning Module Guide.* EPA/400/3-91/002, Washington, D.C.

EPA. 1991. *Lead in School Drinking Water.* EPA 570/9-89-001, Washington, D.C.

EPA. 1991. *Sick Building Syndrome.* Indoor Air Quality Fact Sheet #4, Washington, D.C.

EPA. 1993. *Lead in Your Drinking Water.* EPA 810-F-93-001, Washington, D.C.

EPA. 1995. *Citizen's Guide to Pest Control and Pesticide Safety.* EPA-730-K-95-001, Washington, D.C.

EPA. 1995. United States Environmental Protection Agency and the United States Consumer Product Safety Commission Office of Radiation and Indoor Air (6604J). *The Inside Story: A Guide to Indoor Air Quality.* EPA Document # 402-K-93-007, Washington, D.C.

Fanger, P.O. 1988. "Introduction of the Olf and Decipol Units to Quantify Air Pollution Perceived by Human Indoors and Outdoors." *Energy and Buildings* 12:1-6.

Fanger, P.O. 1989. "A New Comfort Equation for Indoor Air Quality." *Proceedings of IAQ '89: The Human Equation: Health and Comfort.* American Society of Heating, Refrigeration and Air-Conditioning Engineers, Atlanta.

Hansen, Shirley J. 1991. *Managing Indoor Air Quality.* Fairmont Press, Lilburn, Ga.

Department of Housing and Urban Development (HUD). 1985. *The Noise Guidebook.* HUD-953-CPD, Office of Community Planning and Development, Washington, D.C.

Illuminating Engineering Society of North America (IES). 1981. *1981 IES Lighting Handbook: Application Volume.* Waverly Press, Baltimore.

International Conference of Building Code Officials (ICBA). 1988. *Uniform Building Code.* Whittier, Calif.

Internet, Encyclopedia Brittannia: http://www.eb.com:195/cgl-bin/g?DocF=macro/5005/71/115

Lodge, James. 1989. *Methods of Air Sampling and Analysis,* 3rd ed. Lewis, Chelsea, Mich.

Morey, P., J. Feeley, and J. Otten. 1990. *Biological Contaminants in Indoor Environments.* American Society for Testing and Materials Publications, Philadelphia.

National Research Council (NRC). 1981. *Indoor Pollutants.* National Academy Press, Washington, D.C.

Southern Building Code Congress (SBCC) International. 1986. *Standard Building Code.* Birmingham, Ala.

Sheet Metal and Air Conditioning Contractor's National Association, Inc. (SMACNA). 1988. *Indoor Air Quality.* Vienna, Va.

Public Health Service, Office on Smoking and Health. 1986. *The Health Consequences of Involuntary Smoking. A Report of the Surgeon General.* Atlanta.

Public Health Service. Centers for Disease Control. 1991. National Institute for Occupational Safety and Health. Current Intelligence Bulletin 54: *Environmental Tobacco Smoke in the Workplace—Lung Cancer and Other Health Effects.* DHHS (NIOSH) Publication No. 91-108. Atlanta.

World Health Organization (WHO). 1983. *Indoor air pollutants. Exposure and Health Effects Assessment.* Euro-Reports and Studies No. 78. World Health Organization Regional Office for Europe, Copenhagen: WHO-1983.

WHO. 1984. *Indoor Air Quality Research.* Euro-Reports and Studies No. 103. World Health Organization Regional Office for Europe, Copenhagen: WHO-1984.

WHO. 1987. *Air Quality Guidelines for Europe,* European Series No. 23. Copenhagen.

Williams, Phillip, and Robert Greene. 1996. *Indoor Air Quality Investigation Protocols. J. Environmental Health,* October 1996, pp. 6–13.

Walkinshaw, D.S. 1988. The Sick Building Syndrome. Presentation to Ontario Association of School Business Officials' Operations, Maintenance and Construction Workshop: Indoor Air Quality in Schools, Kitchener, Ontario.

Chapter 2

What Factors Can Affect an Indoor Environmental Quality Complaint?

PHILIP HAGAN, MPH, CIH, CHMM

INTRODUCTION

This chapter examines the factors that the building manager should consider when a general indoor environmental quality (IEQ) complaint is made, including the multiple potential causes of IEQ problems.

WATER QUALITY

Occasionally, occupants will voice concerns about the quality of the building's drinking water—especially regarding taste, appearance, and chemical and biological contaminants. How should a building manager address these concerns?

Building occupants will request that their drinking water be tested to determine if it is "safe" to drink. Any test performed on a water sample collected at one point (i.e., grab sample) in your water system will provide information only on the water tested at that particular location and time. This type of testing should be performed only if in-house contamination is suspected.

Water supplies in the United States used for drinking must meet certain Environmental Protection Agency (EPA) standards in accordance with the Safe Drinking Water Act (SDWA). The SDWA requires that suppliers of drinking water test for 84 contaminants (see Appendix H for a complete listing) to ensure that the water delivered to the tap meets these contaminant specifications. Often, states have developed their own guidelines that are more restrictive than EPA standards. Contact your state EPA to determine if this is the case. You can contact your water supplier and ask if your water meets federal guidelines. Federal law also requires consumers to be informed if violations of drinking water regulations occur. If building occupants are concerned about water quality, test results can be obtained from the local water authority. The EPA will answer questions concerning drinking water via its Safe Drinking Water Hotline (800-426-4791).

Bottled-water suppliers do not have to meet same federal guidelines or standards required by the SDWA. However, on May 13, 1996, new bottled-water regulations from the Food and Drug Administration (FDA) took effect. These regulations were targeted primarily at alleviating consumer confusion about the many different types of bottled water on the market by providing standard definitions and bringing mineral water under existing quality standards for bottled water. (Mineral water had previously been exempt from standards that applied to other bottled water.) If your building occupants use bottled water, you should determine what testing is done and what mechanism is in place to inform you if there is a problem.

Bottled water, like all other foods regulated by the FDA, must be processed, packaged, shipped, and stored in a safe and sanitary manner with labels that are both truthful and accurate. Bottled-water products must also meet specific FDA quality standards for contaminants. The following definitions have been established by FDA for the different types of bottled water:

Artesian water or *artesian well water:* Water from a well tapping a confined aquifer in which the water level stands at some height above the top of the aquifer.

Bottled water: Water that is intended for human consumption and is sealed in bottles or other containers with no added ingredients except that it may contain safe and suitable antimicrobial agents.

Ground water: Water from a subsurface saturated zone that is under a pressure equal to or greater than atmospheric pressure.

Mineral water: Water containing not less than 250 parts per million total dissolved solids, originating from an underground water source. No minerals may be added to this water.

Purified water: Water that has been produced by distillation, deionization, reverse osmosis, or other processes and that meets the definition in the United States Pharmacopoeia, 23d Revision, January 1, 1995.

Sparkling bottled water: Water that, after treatment and possible replacement of carbon dioxide, contains the same amount of carbon dioxide that it had at emergence from the source.

Spring water: Water derived from an underground formation from which water flows naturally to the surface of the earth.

Sterile water or *sterilized water:* Water that meets the requirements under "Sterility Tests" in the United States Pharmacopoeia, 23d Revision, January 1, 1995.

Well water: Water from a hole bored, drilled, or otherwise constructed in the ground that taps the water of an aquifer.

The regulation also addresses various other labeling concerns. For example, water bottled from municipal water supplies must be clearly labeled "from a community water system" or, alternatively, "from a municipal source," unless it is processed sufficiently to be labeled as distilled or purified water. The regulation also requires accurate labeling of bottled water marketed for infants. If a product is labeled "sterile" it must be processed to meet FDA requirements for commercial sterility (see definition). Otherwise, the labeling must indicate that it is not sterile and should be used to prepare infant formula only as directed by a physician or according to infant formula preparation instructions (*Federal Register,* November 13, 1995).

The International Bottled Water Association published a model bottled water regulation in 1985 that has been used as a basis for new bottled-water regulations in Arizona, California, Connecticut, Florida, Hawaii, Louisiana, Maryland, Massachusetts, New Hampshire, New Jersey, New York, Ohio, Oklahoma, Pennsylvania, Texas, and Wyoming. If your state does not regulate bottled water and you have specific questions on that topic, the International Bottled Water Association will provide information by phone (703-683-5213).

People frequently voice concern as to why chlorine is added to drinking water. The reason is to kill microbial contaminants. Chlorine is most effective against disease-causing (pathogenic) bacteria. Before chlorine disinfection was instituted, cholera and dysentery were epidemic in the United States. The use of chlorine disinfection to control waterborne diseases is generally considered a public health success story. However, chlorine is not effective against viruses and protozoa organisms. And to further complicate matters, chlorine can react with naturally occurring organic substances in water to form disinfection by-products. Presently, the EPA regulates one of these classes of disinfection by-products called trihalomethanes (THMs). If THMs are causing a problem in your water, you will receive notification from your water supplier. Alternate methods of disinfection (ozone, chloramines, chlorine dioxide) are being used in Europe and some parts of the United States. At this time the potential side effects are not as well known as those associated with chlorination, and more research is needed before these substances will be accepted as practical alternatives.

Because water suppliers are not able to test for every possible pathogen that can occur in drinking water, often other parameters are used as indicator organisms. Coliforms are one class of microorganisms frequently used as an indicator organism, the presence of which indicates a problem. Detection of high levels of coliforms in drinking water frequently result in boil-water orders. Coliforms do not necessarily cause health problems; however, close attention should be paid to the presence of one type of coliform, *Esherichia coli* (*E. coli*), which can cause illness. Luckily, outbreaks of illness due to this organism in drinking water are rare.

Carbon filters meeting National Sanitation Foundation (NSF) Standard 53 are effective in removing cryptosporidium, a protozoan parasite. An occurrence of cryptosporidium in the municipal water supply in Milwaukee in 1993 resulted in severe health problems in the population. The municipal water supply in Washington, D.C., was also found to be contaminated with cryptosporidium resulting in a boil-water order from city officials. Cryptosporidium can be a severe hazard to individuals with compromised immune systems.

Water should be sampled for the presence of lead. Although lead pipes have not been used in plumbing work for decades, lead solder was used to join pipes together until 1986. In addition, water fountains can contain lead-lined water reservoirs. The test for lead is relatively inexpensive and easy to perform. Your local health department can provide information on how to conduct the test as well as names of laboratories or consultants who can perform the test.

If a water test indicates that the drinking water contains lead above 15 parts per billion (ppb), then you should take the following precautions. Since the highest concentrations of lead occur in water that has been sitting in contact with pipes overnight (longer than six hours), running the water for several minutes before use will minimize potential exposure. However, in a high-rise building running the water before use may not lessen your risk from lead. Those plumbing systems have more, and sometimes larger, pipes than smaller buildings do. In this case, you should try to locate the source of the lead; contact a consultant for advice on reducing the lead level. Hot water from the tap is another potential source of lead exposure because of older pipes. A simple solution is to drink and cook with cold water. Hot water can be safely used to wash dishes.

Sometimes drinking water appears discolored. One cause of discoloration is the presence of iron. When iron is dissolved in water it is colorless. However, when it combines with air as the water comes from the faucet, it turns reddish brown. Sometimes brown water is due to rust in the supply pipes or a rusting water heater. If this is a localized problem, then the rust will be found close to the water source. Black water can be caused by dissolved manganese, which combines with chlorine and turns black. If you experience these problems and suspect they are not caused by your building, contact your local water authority for information. Under normal conditions, the contaminants causing these types of discoloration are considered nontoxic. As a prudent measure, a building manager should confirm the reasons for unusual odors or discoloration in the building water supply.

NOISE

Put simply, noise is any unwanted sound. The unit of measure for sound is the decibel (dB), which can be measured by commercially available noise meters. The decibel is used to indicate the relative loudness of sound. The relative value of each decibel measurement increases by factors of 10. A 30-decibel sound is 10 times louder than a 20-decibel sound, and a 40-decibel sound is 100 times louder than a 20-decibel sound. The Occupational Safety and Health Administration (OSHA) regulates exposure to sound in the workplace at 90 dB for an eight-hour exposure period. Very seldom is this regulatory limit exceeded in a nonindustrial workplace.

Normal conversation measures about 30 dB, and typical office noise values range from about 60 to 70 dB. Thunder, a rock concert, an unmuffled large diesel engine, or noise measurements taken close to a freight train could result in levels of 90 to 110 dB.

In the workplace, even noise below regulatory levels can result in IEQ complaints. Luckily, most noise sources can be readily identified, and removing the source is the most obvious solution. Since this is not always possible, increasing the distance between the noise source and the individual who made the complaint will sometimes solve the problem. Other times, the use of acoustical sound barriers can be a successful mitigation measure.

If noise sources are not easily identified, an easy method of locating the source is to shut down noise sources one at a time. While searching for sources of noise, include sources of vibration as well; vibration can produce unwanted noise. Often, vibrating air duct grilles will result in a difficult-to-identify source of noise. If you suspect a vibrating grille, removing it will confirm whether it is the source.

The design of office spaces should incorporate common sense. Background noise increases when sound is reflected from hard surfaces (e.g., bare walls, ceilings, floors, windows, and concrete surfaces). Soft coverings (e.g., curtains, screens, acoustical ceiling tiles, and carpet) can reduce this reflected noise. When modular office systems are installed, partitions should provide reasonable sound attenuation. Noisy equipment should not be placed in areas where it will distract occupants. Eliminating air gaps (e.g., spaces under doors, gaps beneath modular office enclosures) will increase the effectiveness of sound barriers. Walls and floors are assigned sound transmission class (STC) ratings that reflect how well they inhibit airborne sound. The higher the rating, the better the material acts as a sound barrier. When planning renovations, evaluate STC ratings to ensure background noise is kept to a minimum.

In workplaces with machinery there is a potential for exceeding the 90 dB noise level, usually not for the eight-hour period. Any operation that produces power can result in the production of noise. Typical sources of noise in building mechanical systems are as follows (American Society of Heating, Refrigerating, and Air-Conditioning Engineers [ASHRAE], 1993):

- Rotating and reciprocating equipment such as fans, motors, pumps, and chillers
- Air and fluid noises such as those associated with ductwork, piping systems, grilles, diffusers, terminal boxes, manifolds, and pressure-reducing stations
- Excitation of surfaces, for example, friction; movement of mechanical linkages, ducts, and pipes

LIGHTING

For most tasks performed in a building, vision is the main sensory channel for receiving information. As a result, lighting becomes very important when evaluating the indoor environment. Two common terms used to describe lighting are *illumination* and *luminance.* Illumination, measured in foot-candles, is the measure of ambient light falling on a work surface or task. Luminance is a measure of reflected light from the same surfaces or tasks. Lighting should be suited to relevant job tasks; otherwise it can be the cause of IEQ complaints. Too much or not enough light can result in physical distress. Eyestrain, irritation, and headaches are common symptoms that inadequate lighting can cause.

Two basic characteristics of light that can lead to IEQ complaints can be easily assessed: *brightness* and *glare.* Brightness is the relationship between the light supplied to the work surface and the surrounding level of illumination (background light). If the light at a work surface is substantially greater than the surrounding illumination, then the brightness should be reduced.

Direct glare is caused when light in the field of vision is much brighter than light from the working surface or task materials. Reflected glare is caused by the reflection of light from a surface. An easy way to determine if glare is a problem is to place a mirror on the work surface with the worker in his or her normal work position. If the light is reflected into the worker's eyes, then there is too much glare. This can be corrected by moving the light or the work surface.

Glare from reflected light, light shining through windows, or other bright light can create difficulties for individuals using video display units. Using blinds, screens, or awnings; painting walls with nonreflecting paint; or placing workers with their backs to a dark-colored wall will reduce glare. Mirrors, clocks, pictures, or other items that reflect light can be repositioned. Glare screens can be used to reduce glare when working with video display units.

Light supplied from fixtures is usually either *direct, indirect,* or a combination of the two. Fixtures that face down usually provide direct lighting. Indirect lighting is provided by fixtures that are directed toward a wall or ceiling. Advantages of indirect lighting include glare reduction and uniformity of light levels.

Indirect lighting tends to be less efficient (i.e., higher cost) than direct lighting. A combination of the two probably provides the best method of lighting an area.

Color schemes can influence an individual's perceptions in the workplace and can make a workplace more comfortable for occupants. Generally, color preferences range in descending order as follows: blue, red, green, violet, orange, and yellow. Blue, green, and violet are considered cool colors; red, orange, yellow, and brown are considered warm. A person's perceptions of size and distance are thought to be affected by color. The use of blues and greens can make a room appear to be larger, while the use of red can have the opposite effect. Pastel shades tend to diminish the effects of color on a person's perceptions. Before renovating a work space, it is a good idea to get advice from an interior designer or decorator to help maximize the positive effects of a new decor on building occupants.

Sometimes building occupants complaint about inadequate light. The Illuminating Engineering Society of North America (IES) *Handbook, Reference Volume* is a good source of recommended lighting levels for various tasks and space-use considerations. Space-use considerations (e.g., hallways, classrooms, auditoriums, storage areas, office work) generally result in requirements for different levels of illumination. Illumination levels can be determined with a light meter.

EXTERNAL FACTORS THAT AFFECT IEQ: OUTDOOR ENVIRONMENT/CLIMATE/WEATHER

Because of the ongoing exchange (mechanical, natural, infiltration) of air between a building and the outside environment, climate and weather conditions can have a significant impact on the indoor environment. This is due to several factors:

Temperature and sunlight: Particularly for areas where ozone-containing smog is a problem, the temperature and intensity of sunlight can have a strong influence on IEQ. High temperatures typically increase the evaporation rate of volatile organic compounds (VOCs) and semivolatile organic compounds (SVOCs). Energy from sunshine speeds up photochemical reactions that cause smog. This is why smog almost always occurs either in the summer or in southern geographic climates.

Transport of pollutants: Frequently, prevailing winds will transport pollutants to other geographic locations. The long-range and short-range transport effect can be significant for ozone, acid rain, and fine-particulate pollution.

Dilution rate of emitted pollutants by the environment: Usually, emissions from smokestacks or vehicles become mixed with cleaner air, and pollutant concentrations remain low. However, when weather conditions are stable with light winds and little vertical air motion, then the air volume for dilution and the mixing rate are both reduced. As long as such conditions persist, pollutants can concentrate. Vertical air movement is greatly restricted by temperature inversions, scenarios in which the air temperature gets warmer with height instead of cooler. Temperature inversions can concentrate pollutants near the ground and result in their reentry into a building. Local weather stations and airports can provide information as to whether this weather phenomenon is occurring. Temperature inversions can also cause concentrated pollutants to travel some distance from their source and affect buildings throughout a neighborhood. This is another good reason for a building manager to have identified and characterized potential sources of pollutants in surrounding areas.

Another situation that may lead to pollutant buildup occurs when emissions are caught up in air circulating near a coastline or mountains, usually during the summer months. The polluted air may travel for some distance and then return, possibly more than once, leading to adverse air-quality episodes.

Humidity can also affect building occupants' comfort. Humidification of air could be required in colder climates during the winter and dehumidification could be required during the summer in warm climates.

The heating, ventilating, and air-conditioning (HVAC) system is usually the primary driving force behind the movement of pollutants into a building. But wind and differences of temperature between the inside and the outside of a building can overwhelm an HVAC system. Wind tends to act in a transient manner, causing high and low-pressure on different sides of a building. These transient pressure surges can affect exhaust and supply sides of the HVAC system. When complaints of odors come and go and are attributable to no obvious inside pollution source, wind surges could be carrying odors from outside (e.g., vehicle exhaust, garbage) into the building.

The movement of air through currents caused by *convection* can result in the concentration of pollutants. Convection is the tendency for warm air to rise and cold air to drop. Air currents caused by convection within a building can result in a phenomenon known as *stack effect.* Stack effect occurs in multilevel structures when heated air escapes from open windows in higher levels. That air is replaced by air from lower levels, and cooler outdoor air rushes in to replace the heated air that has moved up. This can result in the movement of contaminants from outside to inside and then throughout the building via various pollutant pathways (e.g., elevator shafts, stairwells, plenums, pipe chases, spaces between walls).

Seasonal changes can result in increased pollen loads in the outdoor air. Local media provide area pollen index reports; often an individual's complaints of allergy problems can be traced to the ambient conditions. Hay fever sufferers may first experience symptoms when the pollen count is 10 to 25 (ASHRAE, 1993). Pollen from trees begins creating a problem in April, followed by grasses through the summer and ragweed in late summer. The building IEQ coordinator can record this type of information in a logbook that could be used to correlate health complaints to conditions in the outdoor environment. If someone is experiencing allergic reactions in response to high levels of outdoor pollen, providing information that the building's HVAC filters are decreasing inside pollen levels by 10 to 20 percent compared to outside "fresh" air can suggest that the building is "healthy" and that outside exposures are contributing to the adverse reaction. Pollen grains and fragments range from 5 to 100 microns in diameter and are more readily removed by filtering processes than other dust particles prevalent in outdoor air (0.1 to 10 microns). The building ventilation system can provide cleaner air inside because of the use of particulate filters. This would be good to publicize if it is the case.

A pollutant standard index (PSI) developed by the EPA and the South Coast Air Quality Management District of El Monte, California, has been in use nationwide since 1978 (see Table 2–1). The PSI provides information on air-pollutant concentration and associated health effects.

PSI information is usually available on a daily basis from local newspapers, radio and television stations, and the Internet. Smog concentrations usually peak during the evening rush hour, from about 4 to 7 P.M.

To minimize IEQ complaints, the IEQ coordinator should observe external factors that could impact the indoor environment and ensure that the information is made available to building occupants. When possible, the IEQ coordinator

Table 2–1.   Pollution standard index

Pollutant Standard Index	Health Effects	Cautionary Status
0	Good	
50	Moderate	
100	Unhealthful	
200	Very unhealthful	Alert: Elderly or ill should reduce activity and stay indoors
300	Hazardous	Warning: General population should reduce physical activity and stay indoors
400	Extremely hazardous	Emergency: No physical exertion, everyone stay indoors
500	Toxic	Significant harm: Everyone stay indoors

should also initiate response measures to minimize impact on the building population.

INTERNAL SYSTEMS FACTORS: POLLUTANT PATHWAYS AND DRIVING FORCES

When trying to mitigate causes of an indoor air quality (IAQ) complaint attributed to a contaminant in the work space, two questions have to be answered:

1. What was the driving force used to put the contaminant in the complaint area?
2. What path did the contaminant use?

A simple example would be a housekeeper carrying a pail of floor stripper (contaminants) into an occupied space. The driving force would be the housekeeper and the path would be the doorway of the room.

Pressure relationships are usually the driving force behind contaminant movement. The distribution of air in a building is regulated by pressure relationships. If more air is supplied to an area than is removed by the HVAC system, then *positive pressure* exists. Positive pressure results in leakage of air from an area into adjacent spaces (e.g., filling a balloon with air will result in positive pressure in the balloon compared to its surroundings; the air is pushing on the inside of the balloon trying to escape). If pollutants are in the area of positive pressure, they can be pushed into adjacent areas.

If less air is supplied to an area than is removed by the HVAC system, air tends to leak into the space, creating what is known as *negative pressure.* Pollutants in adjacent spaces will be drawn into the area of negative pressure.

These pressure relationships are the driving forces behind the movement of pollutants from one location to another (e.g., if there is a hole in the balloon, air will push out into the surrounding space). These driving forces move pollutants through any accessible openings. For example, a building with odorous activities in one area (e.g., a beauty salon) may have its HVAC system designed so that airflow-pressure relationships keep the odors away from other spaces; but if a worker props open an emergency exit door or a building occupant opens a window, the pressure relationships may change so that the odors travel to previously protected areas, resulting in IAQ complaints.

These accessible openings are also known as *pollutant pathways.* Pollutants use air pathways to travel from one location to another. These pollutant pathways are governed in large part by building HVAC systems. HVAC systems provide the pollutant pathways (duct systems). However, the potential distribution of contaminants in a building can be affected by any of the system components in a building. Pollutant pathways are provided by windows, doors, ducts, spaces between walls, suspended ceilings, elevator shafts, stairwells, pipe chases, dumbwaiters, and ceiling plenums. Since buildings are designed to use HVAC systems to move air from one location to another, HVAC systems are always a potential source of investigation when the source of an air-quality problem is not readily identified.

Elevators are frequently found to provide both a pollutant pathway and a driving force for pollutant movement. As an elevator moves up and down its shaft (pollutant pathway), a piston effect (driving force) is created, potentially moving pollutants throughout a building. To compound this effect, often either chemicals are stored in elevator rooms or in areas that provide cross connections to machinery spaces resulting in sources of indoor contaminants.

HEATING, VENTILATION, AND AIR-CONDITIONING SYSTEMS

Most people tend to think of ventilation as either air movement in a building or the supply of outdoor air to a building. Actually, ventilation is a combination of processes that results in the supply, conditioning, mixing, return, and removal of air from inside a building. When events serve to alter one or more of these processes, the quality of indoor air may deteriorate, leading to occupant discomfort or exposure to hazardous substances.

HVAC systems are designed to supply air to spaces in buildings. Air-conditioning provides occupants with thermal comfort. The conditioned air should provide an environment with "comfortable temperature and humidity levels, free of harmful concentrations of air pollutants . . . ventilation is actually a combination of processes that results in the supply and removal of air from inside a building. These processes typically include bringing in outdoor air, conditioning and mixing the outdoor air with some portion of indoor air, distributing this mixed air throughout the building, and exhausting some portion of the indoor air outside" (US EPA, 1990).

HVAC systems should be professionally balanced to provide pressure relationships that minimize the migration of pollutants to occupied spaces. Building managers should monitor and prevent unauthorized openings of doors and windows that can defeat the value of the air-handling design.

Air is often recirculated back to occupied spaces as an energy-conservation measure. The recirculated air is mixed with supply air and supplied back to the occupied spaces. Recirculated air should be filtered to ensure that the air is clean. Systems should be maintained to minimize the dispersion of odors and other contaminants. Replace filters according to the manufacturer's recommendations.

Sound design and operation of HVAC systems, coupled with good maintenance practices, can eliminate many potential IAQ problems. When budgets
must be cut back, money slated for preventive maintenance is often the first item
to fall by the wayside. It is true that this saves money in the short term, but defer-
ring maintenance will result in increased future costs and potential dissatisfaction of employees or customers.

Typical HVAC systems have several common features: controls (on-off
switch and thermostat), an air mover (fans), a distribution network (ducts), and a
cleaning and conditioning system (filters and cooling/heating components). A
regular preventive maintenance schedule should include each of these components. Building occupants should be educated as to how the air in the building is
conditioned and how the limitations and parameters of the operating system
could affect their work space.

The three general categories of HVAC systems are *central systems, perimeter systems,* and *unitary systems.*

Central Systems

Central systems are large, centrally located HVAC systems that serve many
building zones through an enclosed air-distribution system. Typically, ducts deliver preconditioned air from a central location(s) throughout zones served by the
air-distribution system. Most central systems are of two general types: *constant
air volume* or *variable air volume.* These two central-system HVAC designs differ in that one uses constant air volume at different temperatures and the other
uses variable air volume at a constant temperature to control the indoor climate.

Constant air volume systems control temperature (heating and cooling) load
by varying the temperature of constant volume supply air. The percentage of outside air can be held constant or controlled to vary with outside temperature and
humidity. Conditioned air is usually supplied to zones through duct distribution
systems. Often, temperature can be further controlled in individual zones by the
use of reheat coils located in ducts or bypass lines that send the supply air back
into the return air before it reaches the work space.

Variable air volume (VAV) systems supply air at a constant temperature to
VAV distribution boxes located throughout the various zones in a building envelope. Temperature (usually cooling) control is achieved by varying the amount
(volume) of supply air delivered to building zones. When VAV systems are used,
perimeter zones (located on the outside of a building envelope) are usually
heated by another kind of system. With early VAV designs, outside air supplies
were not controlled, causing a subsequent decrease in the supply of outside air as
airflow decreased in response to building temperature requirements. Often, when
VAV systems have achieved the desired temperature setting for a space, the volume of outside air supply can go to zero. The reduction in outside air translated

into energy cost savings by reducing the amount of air that had to be conditioned.

As a result of the energy cost savings, VAV systems have become popular in recent years as building managers have sought to implement energy-saving measures. Later designs have incorporated minimum quantities of outside supply air through the use of static pressure devices in the supply airflow. If complaints are received concerning indoor air quality under these conditions and the lack of outside air is identified as the culprit, usually the system can be reengineered to supply a percentage of outside air to the space at all times.

Perimeter Systems

Perimeter systems, used in combination with other systems or by themselves, condition air in zones located along the perimeter of a building. Typical perimeter systems are designed to handle changes in outer zones that are influenced by outside weather conditions. Air-water induction units, electric baseboard heaters, fan-coil units, and fin-tubed radiators are examples of perimeter systems.

Unitary Systems

Unitary systems are premanufactured HVAC systems. Usually each zone is served by a separate unit. Rooftop units, through-the-wall conditioner systems, and heat pump systems are examples of unitary systems.

Potential Problems with HVAC Systems

Some systems operate with intermittent airflow as an energy-conservation measure. This can lead to buildup of indoor contaminant levels, as it prevents the HVAC system from removing contaminants. Ventilation operating parameters should ensure prevention of contaminant buildup. Reduced or interrupted flow schedules should be examined to ensure sufficient outside air exchange. HVAC operating parameters should always take into account building-use schedules (e.g., 8-hour operations as opposed to 24-hour operations).

By design and function, HVAC systems deliver conditioned air to the interior of buildings. If an HVAC system is operating according to design parameters and the building use is in keeping with that design, thermal conditions should satisfy the majority (at least 80 percent) of building occupants. Typical design parameters for HVAC systems are for operating ranges of levels within which variation is acceptable. As long as the conditioned air provides building occupants with a satisfactory thermal environment, the distribution system is working.

Building occupants who become dissatisfied with thermal conditions may respond in a variety of ways. Some occupants alter air-distribution patterns by covering registers. Some change the setting on local thermostats. Others open windows, move furniture, or prop doors open. Any of these operations can serve to further exacerbate thermal discomfort by affecting pressure relationships throughout the building. It is important to respond in a timely fashion to thermal complaints to minimize occupants' response actions that could aggravate the problem.

Indoor air quality issues should be addressed throughout the life of a building. During planning stages, HVAC systems should be designed to provide minimum outside ventilation rates and acceptable IAQ that minimizes the potential for adverse health effects (ASHRAE, 1989). Specialists (e.g., industrial hygienists, mechanical engineers, architects) with experience in IAQ should be consulted. Designs should address the big picture and anticipate future building uses. HVAC systems should condition air, deliver the air to the occupants, and return the air to be either reconditioned or exhausted to the outside according to a specific design. Often, changes made during construction do not take original design considerations into account. This can result in poor IAQ and adverse effects to building occupants.

Locations of air intakes should take into account potential sources of pollutants in the surrounding neighborhood. Building materials should not be used that will result in a pollutant load that the HVAC and associated building systems cannot handle.

Common sense can eliminate potential flaws in the design of new construction or renovations. Some mistakes that appear all too often due to a lack of common sense include:

- Air supply and return air vents located adjacent to each other inside occupied spaces, frequently resulting in short-circuiting of the air supply. The supply air leaves via the return air vent before it has a chance to refresh the occupied area.
- Outside air supply intakes located next to exhaust vents. Quite often this results in reentrainment of discharged contaminants. Contaminants are then delivered back to the building through the HVAC system, increasing pollutant concentrations in the building.
- Outside air supply vents located near pollution sources. Classic examples include locating air-supply vents next to a loading dock, a building garage, a bank or a fast-food drive-up window, or a solid waste or trash-collection area. Wind currents should be examined to ensure that waste-collection areas are located downwind of the air intakes to minimize entrainment of odors.
- Mechanical systems built and installed not in accordance with original blueprints. Frequently, mechanical systems are installed by subcontractors

who are forced to take "detours"—alternate paths or shortcuts necessary because of work completed previously by other contractors. For example, a duct that has to be rerouted around an electrical conduit could add four 90-degree turns, resulting in decreased airflow.

- Change orders for new construction and renovation projects not evaluated for their potential impact on original design parameters for mechanical and plumbing systems. This can result in undersized fans due to the addition of extra ductwork, thermostats placed in the wrong location, walls or partitions placed where airflow is blocked, and other similar problems.

After the occupants are in the new building or after extensive renovation has occurred, problems reported with efficiency and airflows are usually addressed as part of the construction company's final adjustments. Once the building has been in use, there may be a need for postoccupancy evaluation (POE), a technical examination of the actual operating flows within the building. The POE analysis of building use helps the owner understand what additional steps to undertake when the building is retrofitted or later modified.

Building commissioning is the process by which building systems are evaluated by engineers, during design and construction and after construction is completed, to ensure that systems operate in accordance with design parameters. New buildings should be commissioned before inhabitants take up residence. During the commissioning process, the HVAC system should be tested and balanced in accordance with design specifications. The commissioning process should provide documentation and training requirements for operations and maintenance procedures. Documentation and information obtained from the building-commissioning process should be used to evaluate how well the building is functioning throughout the life of the building. ASHRAE Guideline 1-1989, *Guideline for Commissioning of HVAC Systems,* can be used to develop criteria for commissioning practices.

Humidification and Moisture

Problems with humidity are possibly the single largest IAQ problem associated with HVAC systems. If not enough humidity is in the air (greater than 20 percent), people tend to complain of dry mucous membranes. Humidity levels higher than 60 percent can contribute to microbial growth (molds and mildew).

Humidifiers can spray steam or water into the air to increase humidity. The use of steam systems minimizes the risk of microbial contamination. Dehumidification systems remove moisture from the air, but it is very important to conduct regular maintenance on the systems to ensure microbial growth does not occur in the water reservoirs (condensate drain pans). Since cooling coils are used for dehumidification in most systems, care should be taken to ensure that they are maintained properly for maximum efficiency. Humidification and dehumidification systems must be kept clean to prevent the growth of pathogenic organisms (bacteria and fungi). Improperly functioning humidifiers can create areas where ponding of water can occur. This happens when spray wets a surface. Accumulations of water anywhere in an HVAC system may foster harmful biological growth that can be distributed throughout the building. Moisture from condensation, leaks, and spills can foster the growth of microbial contamination.

Condensation occurs when air is saturated with moisture (and cannot hold any more) and the temperature drops. The capacity of air to hold water decreases as temperature falls. Relative humidity values indicate what percentage of water the air can hold at any given temperature. Inside a building, if a surface is colder than the temperature of surrounding air that is saturated with water (100 percent relative humidity), moisture will condense onto the cold surface. This happens because the cold surface lowers the temperature of the surrounding air, decreasing the amount of water that it can hold. Condensation can also occur on the inside of exterior walls that are cooled by winds blowing on the outside of a building.

Condensation can occur on the discharge side of an HVAC outdoor air supply fan due to the large pressure drop that occurs when air moves from the fan into the duct system. If the interior of the duct is lined with porous insulation, conditions (i.e., moisture, food, and temperature) are conducive for microbial growth. When searching for the origin of a microbial growth problem, check this scenario as a potential source.

If cool air from a supply air vent is directed toward a wall, ceiling tile, or other similar surface, water will condense in the area cooled by the vent. This can be corrected by redirecting the vent to change the direction of the supply air. Any time condensation occurs, conditions are probably conducive to microbial growth.

If drip pans or catch basins are used in a building, care should be taken to ensure that proper drainage is occurring and that pans or basins that use evaporation are not serving as reservoirs for microbial growth.

Checking for Common HVAC Ailments

Often problems with an HVAC system can be readily identified. If ventilation seems to be a problem, a number of items can be readily evaluated.

The occupant can check to determine the following:

- Is the thermostat set correctly?
- Is warm air flowing out from the supply vent in the winter?
- Is cold air flowing out in the summer?

Building engineers can check to determine the following:

- Is the ventilation system turned on?
- Are drain pans clean and draining properly?
- Is the outdoor intake or supply vent blocked?
- Is the fan(s) operating?
- Are the motor controllers functioning properly?
- Is the filter(s) clean and properly installed (tight fit)?
- Is there any standing water in the unit?
- Is there any visible microbial contamination?
- Are the coils clean?
- Is the exhaust fan turned on?
- Is air flowing out of the exhaust vent?
- Are the dampers working properly?

Air Cleaning

Typically, air is cleaned through the use of filters in the HVAC system. Different kinds of filters remove particulates and gases.

Particulate cleaners use mechanical filters and/or electrical charges to entrap particles. Mechanical filters physically remove particulates. Electronic methods of removal use differences in electrical charge to remove particulates. Particulates have to be suspended in air before the filter systems can clean the air.

Generally, mechanical filters are rated by efficiency percentages based on one of several standard industry tests. The three tests are as follow:

- Departmenet of Defense Efficiency Percentage Military Standard 282 (DOD)
- Dust Spot Efficiency Percentage ASHRAE Standard 52.1-1992 (Dust Spot)
- Arrestance Percentage ASHRAE Standard 52.1-1992 (Arrestance)

The three tests are not equivalent and the efficiency percentages cannot be compared directly.

DOD-rated filters are considerably better at filtering respirable particles than the other two types. High-efficiency particulate air (HEPA) filters are included in this category. The dust spot and DOD tests measure the number of particles that a filter captures. Put simply, the arrestance test measures the weight of particles that are captured.

The arrestance test method is used mainly to evaluate the performance of filters that help keep the HVAC system operating at peak efficiency. Dust spot efficiencies can evaluate effectiveness in removing air pollutants. Most pollen, mold, and spores are removed by dust spot efficiencies of around 40 percent, whereas for arrestance efficiencies to provide similar removal, the rate would have to be greater than 96 percent. If particulates are found to be a problem, filters can sometimes be upgraded. The more efficient a filter is, however, the more energy is required to push air through the filter. An engineer must determine whether an upgraded filter will still allow the HVAC system to supply proper design airflows.

Most microbial agents are larger than 1 or 2 microns in diameter. Filters with a moderate dust spot efficiency of 50 to 70 percent adequately remove this particulate from the airstream (*Guidelines for Assessment of Bioaerosols in the Indoor Environment*, 1989).

Improperly installed filters will not operate at rated efficiencies. If filters are incorrectly sized or seated improperly in their holders, or if gaskets do not provide an airtight seal, then unfiltered air will bypass the filter, following the path of least resistance.

Electronic air cleaners use positively charged electric fields to create positively energized particles that are then removed from the air by attraction to negatively charged plates. These types of filters can provide a high filtering efficiency at a low pressure drop and not greatly affect airflow rates through the HVAC system. Since the system is electronic, an obvious disadvantage is that it is not suitable for use in areas where high humidity can affect operations. As always, preventive maintenance is important. The plates must be cleaned on a regular basis to prevent the charged particles from passing through the collection area into occupied areas.

Mechanical filters that work with static electricity and charged media filters are two types of electrostatic air cleaners that are not very efficient at removing indoor air pollutants from the air.

Charcoal filters are most commonly used to remove gaseous contaminants from air. Effective-charcoal filters are expensive and tend to cut down on the amount of air supplied by the HVAC system. Charcoal filters will not remove all gases from the air. Retrofitting charcoal filters into HVAC systems is usually very difficult and often not feasible. Pellets impregnated with reactive chemicals such as potassium permanganate can be used in filter media to remove some contaminants through chemical reactions. Eventually these filters become saturated and need to be regenerated or replaced. If gaseous contaminants are a serious problem, source management is probably the most effective means of control.

Filters need to be located downstream from coils and drain pans to ensure that microbial contaminants are removed before air is supplied to occupied areas.

Energy Assessment

Energy and environmental costs can be determined by relating costs of electricity to square feet of occupied space. Since most IAQ complaints are related to temperature and humidity, providing thermal comfort can be a very cost-effective mitigation measure. A productive and happy workforce is the building tenant's goal. Volume II of *The Report to Congress on Indoor Air Quality, As-*

sessment and Control of Indoor Air Pollution indicates that labor costs per square foot may be 10 to 100 times greater than energy and environmental control costs.

HOUSEKEEPING

The IEQ management plan should describe the scope of housekeeping duties conducted in occupied spaces. Operating schedules should be announced in advance to building occupants. Materials used in cleaning operations (e.g., shampooing carpets, stripping and waxing floors, polishing, etc.) should be evaluated for their potential impact on the indoor environment.

Often, housekeeping activities (e.g., vacuuming, dusting) results in high levels of airborne allergens (dust, fibers, microbes, etc.). Evaluate scheduling as well as the sensitivities of building occupants to minimize the impact of these operations on the building population.

Carpets have the potential to contribute to IAQ problems. They can be a source of airborne fibers as a result of aging, excessive wear, and/or chemical exposure from adhesives or cleaning agents and can serve as a breeding ground for microbial populations, especially after flooding has occurred.

Inappropriate use of housekeeping cleaning products can have a significant impact on IEQ. Housekeeping staff will often mix cleaning solutions in higher concentrations than are recommended or with other products to increase cleaning power. This practice can result in hazardous residues left in carpets and upholstery.

The improper use of ammonia, a common cleaning product, can result in adverse effects to the indoor environment. In areas where ammonia cleaning compounds are used, ventilation should be sufficient to prevent the buildup of noxious ammonia vapors. Ammonia cleaners should be used in accordance with manufacturers' instructions and not mixed with other products. Mixing ammonia with lye or sodium hydroxide compounds, caustic substances commonly used to clean clogged drains or ovens, can result in the release of toxic ammonia gas. Ammonia gas causes skin, eye, and respiratory irritations.

Mixing ammonia products with bleach can produce harmful chloramine gas by-products. Chloramine gas enters the body through breathing and can travel deep into the lungs, resulting in serious injury. In some instances, for example, in the removal of mold and mildew, bleach is used to disinfect a surface followed by cleaning with an ammonia product. If the bleach is not thoroughly rinsed from the surface before the ammonia product is applied, the resulting mixture can also produce harmful chloramine by-products.

Combining bleach with other products can also result in significant adverse impacts on the indoor environment. Common household bleach contains chlorine (in the form of sodium hypochlorite), a very reactive ingredient. Sometimes, in an attempt to improve cleaning efficiency, workers mix other products with bleach. If an acidic substance (such as toilet bowel cleaner or rust remover) is used, the sodium hypochlorite can release toxic chlorine gas. Exposure to chlorine gas can cause irritation to the eyes, nose, throat, and lungs.

IEQ Information from Material Safety Data Sheets

Material safety data sheets (MSDS) are standardized informational descriptions of hazardous products or substances supplied by the manufacturer or distributor for the benefit of users and their employees. Federal regulations (i.e., the Hazard Communication Standard: 29 CFR 1910.1200) require manufacturers or distributors of chemicals to provide MSDSs for their products. Information from the MSDSs should be used to evaluate potential hazards.

An MSDS is designed to provide both workers and emergency personnel with the proper procedures for handling or cleaning up the substance. These are of particular use if a spill or other type of accident involving the material occurs. An MSDS should be available for all substances used by the housekeeping or maintenance team.

MSDSs present information in many different formats; however, the general categories of information are the same. An MSDS will always contain the following information:

> *Contact information.* The name, address, and phone number of the manufacturer or distributor of the product will be provided. Trade names and synonyms will be listed, along with a Chemical Abstracts Services (CAS) number. Every chemical has a unique CAS number that can be used to identify the chemical. Details about specific chemical properties are usually included—chemical family, molecular formula, and molecular weight. This information will be followed by a list of the types and amounts of hazardous ingredients. Occupational exposure limits will be listed. (As indicated earlier, these exposure limits should not be used to evaluate exposures for office working environments.)
>
> *Physical data.* The section on physical data will include a description and specific chemical attributes—boiling point, melting point, specific gravity (lighter or heavier than water), vapor pressure (vapor emitted from any substance exerts pressure, e.g., steam from boiling water), pH (acid or base), solubility in water (floats or mixes with water), and vapor density (heavier or lighter than air).
>
> *Potential fire and explosion hazard.* Information will be provided on whether the substance can cause fire and/or an explosion. If so, proper firefighting procedures will be described. If the chemical is known to be

reactive and undergoes forceful chemical reactions when combined with other materials, it will be identified as such. Incompatible materials will be listed.

Health effect information. Toxicity information will describe the degree of health hazard associated with the chemical and whether overexposure results in acute (immediate) or chronic (over time) problems. The MSDS will indicate whether the material is a carcinogen (an agent that causes cancer). Basic first aid measures will be provided.

Proper storage and disposal procedures. Typically this is covered by a very general statement: Observe all federal, state, and local regulations when storing or disposing of the substance. You can call your supplier to find out additional information if needed.

Proper emergency spill and response procedures. The MSDS will describe these procedures and the proper use of protective equipment.

NONROUTINE OPERATIONS

Renovation can contribute to IAQ problems by disturbing existing pollutants or by bringing in new contaminants. Dusts, fibers, and bioaerosols are common pollutants disturbed during renovation activities. Complaints during renovations occur frequently, because building occupants are exposed to odors (VOCs) that are not normally present in the workplace. Common sources of these odors include paint, asphalt, and solvents used during cleanup. Occupant discomfort due to thermal conditions (e.g., cold wind through walls) can increase the number of IAQ complaints made during renovation projects. Contract specifications should detail preventive measures contractors will use when they perform work that could impact the indoor environment. A methodology should also be in place to verify that the preventive measures have in fact been implemented and are working.

Concentrations of dusts and fibers can increase during renovation as a result of sanding, grinding, cutting, and demolition. Although some of these are "nuisance dusts" and can be treated as such, care should be taken to ensure occupant exposure to toxic substances is controlled.

Nuisance dusts and bioaerosols can be controlled through the use of temporary barriers, ventilation, scheduling, and housekeeping. Although most dusts encountered during renovation and construction act as irritants, dust generated when lead paint is disturbed can result in toxic exposures. If lead is present, actions such as sanding, grinding, sawing, drilling, and demolition could result in adverse exposures. Paint manufactured before 1950 often contained greater than 50 percent lead. Painted surfaces in buildings built before 1978 should be assumed to contain lead unless industrial-hygiene surveys have proven otherwise. Whenever renovation will disturb painted surfaces in structures built prior to 1978, surveys should be conducted beforehand to determine if lead is present.

The most common fibers disturbed during renovation that result in adverse impacts on IAQ are *asbestos* and *fiberglass*. Asbestos is a natural substance that has been used for decades in buildings for fireproofing, thermal and acoustical insulation, vinyl asbestos floor tiles (VAT), ceiling tiles, acoustical surfaces, mastic adhesive, and roofing materials. Surveys to determine the presence of asbestos-containing materials should be performed before any construction activity takes place. If asbestos-containing materials are present, then plans for safe removal should be included in the renovation design.

Fiberglass is used as thermal insulation and to line ducts as an acoustical control measure. When fiberglass is disturbed, fibers are released, causing eye, skin, and upper-respiratory irritation. Fiberglass can also act as a reservoir for microbial contamination. As with nuisance dust, exposure to fiberglass can be controlled through the use of temporary barriers, ventilation, scheduling, and housekeeping.

Material safety data sheets should be reviewed for any materials used during renovation and any recommended protective measures implemented. VOCs can be controlled through the use of negative air pressure and isolation of affected HVAC systems.

Occupants should receive prior notification of any impacts the renovation process may have on their working spaces. Someone familiar with all aspects of the renovation project should be available throughout the process to address concerns as they arise.

If portable ventilation equipment or barriers are used to control exposure from the work area, care should be taken to ensure pollutant pathways are not created that could funnel pollutants to occupied areas. Renovation projects frequently impact HVAC systems and their intended function in a building. As a result, it is always prudent to investigate HVAC systems when there are obvious signs (odors) of pollutants and no visible sources in the immediate area of the complaint. Before the building population moves back into a renovated space, good housekeeping and ventilation should be used in the area to clear it of contaminants.

SITE CHARACTERIZATION

Urban settings can result in unique problems because of space limitations and close proximity of neighbors. When buildings are constructed very near each other, the air exhaust from one building can become the air supply for another. Heavy traffic in the area can result in elevated carbon monoxide levels and particulate loads during morning and evening rush hours. In this situation, HVAC system operation should be scheduled accordingly. A higher percentage of recirculated air during rush hours could be one response to minimize the intake of carbon monoxide and exhaust fumes.

Typical IEQ problems encountered in residential settings are radon; asbestos; pesticides; combustion by-products from heating appliances; and allergens from dust mites, arthropods (e.g., cockroaches), pets, and other biological sources. These are discussed later in this chapter.

In suburban settings, pesticides and herbicides used in landscaping can be potential sources of pollution migration into an indoor environment. Integrated pest management (IPM) techniques can reduce the amounts of chemical pesticides used in landscaping. IPM techniques are discussed later in this chapter.

Suburbs sometimes have higher smog levels than other areas. This should be taken into account when health complaints are received. In industrial settings, pollutants can come from manufacturing operations, increased traffic patterns, and stack discharges (see "Heating, Ventilation, and Air-Conditioning Systems," earlier in this chapter.) Combinations of urban, residential, commercial, and suburban settings can result in all the problems discussed here.

RESOLUTION OF THE COMPLAINT

As discussed earlier, most IEQ complaints are solved during the initial contact and follow-up. In more complex situations that involve the health of one or more individuals, resolution can be difficult. Sometimes resolution can only occur after a physician has been able to evaluate and treat the individual successfully.

This medical evaluation and follow-up may involve fixing something that is wrong in the indoor environment, or it may not. An important fact often lost in attempts to solve IEQ problems is that all people at some point get sick. Fixing a building will not cure an individual if an inaccurate diagnosis has been made and the building did not cause the health problem.

INDOOR AIR POLLUTANTS

Tobacco Smoke

Environmental tobacco smoke (ETS), or *secondhand smoke,* is the combination of smoke exhaled by smokers and the smoke that comes from the burning end of cigarettes, pipes, or cigars. Research has identified more than 4,000 compounds as combustion by-products from ETS. Forty of the compounds are carcinogens and many others are irritants. Exposure to ETS is often referred to as *passive smoking.*

No general levels of acceptable ETS have been agreed upon. ASHRAE Standard 62-1989 specifies dilution with fresh air at 60 cubic feet per minute per person in rooms where smoking occurs.

Prohibition of smoking in public spaces is a common control measure. Natural ventilation through open windows or the use of exhaust fans can reduce but not eliminate exposure. The 1986 Surgeon General's report concluded that "physical separation of smokers and nonsmokers in a common area may reduce but will not eliminate non-smokers' exposure to environmental tobacco smoke."

If elimination of smoking indoors is not possible, ventilation should be increased in accordance with ASHRAE guidelines. Energy costs associated with upgrading ventilation for smoking areas can result in substantial financial expense. Generally, the most cost-effective way to reduce exposure to ETS is to eliminate smoking.

For publications about ETS, contact EPA's Indoor Air Quality Information Clearinghouse (IAQ INFO), 800-438-4318 or 202-484-1307.

Volatile Organic Compounds (VOCs) and Semivolatile Organic Compounds (SVOCs)

Organic compounds are chemicals that contain carbon. Organic chemicals are found in cleaning products, cosmetics, adhesives, combustion processes, gasoline, paints, pesticides, solvents, tobacco smoke, and personal-care products (scents and hair sprays). Home-care products (finishes, rug and oven cleaners), paints and lacquers (and their thinners), paint strippers, pesticides, dry-cleaning fluids, building materials, home furnishings, office equipment (some copiers and printers), office products (correction fluids and copy paper), graphics and craft materials (glues and adhesives), permanent markers, and photographic solutions are further examples of organic chemicals.

Many of these products can release vapors at normal room temperature and pressure. Those that do are commonly referred to as *volatile organic compounds* (VOCs) or *semivolatile organic compounds* (SVOCs). The scientific definition includes any organic compound that participates in atmospheric photochemical reactions. By definition, VOCs release vapors more readily than SVOCs. Hundreds of VOCs and SVOCs can be found in indoor air; some of them may have short- or long-term health effects.

The EPA conducted Total Exposure Assessment Methodology (TEAM) studies in six communities in various parts of the United States and found that levels of about a dozen common organic pollutants were up to 10 times higher inside homes than outside, regardless of geographic location. Additional TEAM studies indicated that very high pollutant levels can occur when people use products that release VOCs or SVOCs. The studies showed that elevated concentrations persisted in the air long after the activity was completed.

The ability of organic chemicals to affect health varies greatly—some are highly toxic, others have no known health effects. As with other pollutants, the extent and nature of the health effect depend on many factors, including level of exposure and length of time exposed. Eye and respiratory-tract irritations, headaches, dizziness, visual disorders, and memory impairment are among the immediate symptoms that some people have experienced soon after exposure to some organic chemicals. At present, not much is known about what health effects occur from the levels of organic chemicals normally found in homes. Many organic compounds are known to cause cancer in animals; some are suspected of causing, or are known to cause, cancer in humans (US EPA Consumer Product Safety Commission, 1995).

At room temperature, volatile organic compounds are emitted as gases from some solids or liquids. Concentrations of many VOCs are consistently higher indoors than outdoors (US EPA, 1987).

Though the National Institute of Occupational Safety and Health (NIOSH) has recommended occupational standards for many compounds, no standards have been set for VOCs or SVOCs in nonindustrial settings. Typical indoor VOC concentrations are lower than the standards NIOSH recommends. VOCs are typically higher in a building after new furnishings have been installed, or the space has been painted or renovated. VOCs by their nature (volatile) will usually off-gas in about six to eight weeks. *Bakeout* and *flushing* techniques have been used in an attempt to accelerate this off-gassing process. Bakeout uses elevated temperatures in a building to increase the off-gassing rate. Experience seems to indicate this decreases VOC concentrations for short periods before the levels return to prebakeout values. Flushing introduces the maximum amount of outside air into the space until VOC levels decrease. A major disadvantage is the energy cost of conditioning the outside air.

VOCs are ubiquitous in the environment. Low levels of VOCs can be found in all indoor environments. It is difficult to quantify exposures to VOC concentrations, and measurement can be very complicated and expensive. Generally, sampling for VOCs should be avoided unless there is a specific reason for doing so—for example, complying with a request from a treating physician. As with most measurements of chemicals that have not been extensively evaluated, synergistic processes could be occurring, and a knowledgeable physician should evaluate exposures.

Painting operations can be a significant source of VOCs and SVOCs. Latex paints result in lower levels of VOCs and SVOCs compared to oil-based paints. Although VOCs from most paints decrease significantly within 48 hours after application, certain practices help minimize exposure. Identify pollutant pathways and ensure they are not allowing VOCs to migrate to occupied areas. Ventilating the air in newly painted areas to the outside will help prevent this from occurring. Roller and spray paint-application methods produce the most aerosols and have the highest potential for producing VOCs and SVOCs. Containers of paints and other VOC-producing materials should be covered as much as possible. Often when painting is completed, leftover materials are stored in mechanical and elevator rooms. This should not be done. These areas tend to provide pollutant pathways (e.g., ventilation ducts, elevator shafts) to other parts of the building. Leftover paint materials should be disposed of properly and in accordance with local regulations and guidelines.

Where practical, excessive uses of materials containing VOCs should be restricted. Adequate ventilation should always be used with products that off-gas VOCs. Selection of fast-drying materials can result in high VOC concentrations for a short time, but they tend to dissipate quickly. Materials that can produce VOCs should be stored away from occupied areas and potential pollutant pathways.

Formaldehyde

Formaldehyde is a colorless water-soluble gas. Due to its wide use, it is frequently considered separately from other VOCs.

Formaldehyde is one of the more well-known indoor air pollutants that can be readily measured. If formaldehyde is the potential cause of the problem, identify, and if possible, remove the source. If removal is not possible, reduce exposure: use polyurethane or other sealants on cabinets, paneling, and other furnishings. To be effective, any such coating must cover all surfaces and edges and remain intact.

Materials containing formaldehyde are used in buildings, furnishings, permanent-press fabric, mattress ticking, and many consumer products. Formaldehyde-based resins are used in the manufacture of plywoods, particleboard, textiles, and adhesives. Urea-formaldehyde (UF) resins are commonly used in interior-grade plywood and pressed-wood furniture, cabinets, and shelving. The walls of some buildings have been insulated with urea-formaldehyde foam insulation (UFFI). Phenol-formaldehyde (PF) resins are normally used in exterior-grade products. Formaldehyde out-gasses from the products just mentioned. UF-based products typically emit higher levels of formaldehyde than PF-based materials. Tobacco smoke and other combustion products are secondary formaldehyde sources. Sensitive individuals may choose to avoid these products.

OSHA has set a federal standard for occupational exposure to formaldehyde, a permissible exposure level (PEL) of 0.75 ppm and an action level of 0.5 ppm. OSHA regulates formaldehyde as a carcinogen. The agency also requires that labels be posted in workplace entrances informing exposed workers about the presence of formaldehyde in products that can cause levels to exceed 0.1 ppm. Some states have established a standard of 0.4 ppm in their codes for residences; others have established much lower recommendations (e.g., the California guideline is 0.05 ppm). Most people can detect formaldehyde at levels of about 0.1 ppm. Based on current information, it is advisable to take actions, such as ventilation, to mitigate formaldehyde that is present at levels higher than 0.1 ppm.

Increased temperature and humidity will accelerate outgassing of formaldehyde. Therefore, ventilation may not be an effective means for mitigation. Some manufacturers are producing products with lower outgassing rates. Some surface treatments are being used to seal against outgassing, but long-term effectiveness is still not known. Indications are that outgassing from materials diminishes over time (this could take up to two years).

Pest Control and Pesticides

Pesticide is a general term for products used to kill or control pests, which include plants, insects, rodents, bacteria, fungi, and weeds. Cockroaches, mice, rats, mold, and mildew are common sources of allergens and are frequently controlled by the use of pesticides. The EPA regulates the use of pesticides under the Federal Insecticide, Fungicide, and Rodenticide Act (FIFRA). Pesticides sold in the United States must be registered by the EPA and bear an EPA registration number.

Pesticides are used to control insects (insecticides), termites (termiticides), rodents (rodenticides), fungi (fungicides), and microbes (disinfectants). Pesticides have been identified as an indoor air pollutant in some IEQ building investigations as a result of their widespread use in controlling building pests, including those associated with food-preparation areas, indoor plants and living spaces and those that are tracked in from the outdoors. Pesticides are toxic to specific organisms. They can be highly toxic and short-lived in the environment, or less toxic and very persistent. Persistence in pesticides ranges from moderately persistent (a lifetime of 1 to 18 months—e.g., 2,4-D—atrazine); to persistent (lifetime up to 20 years—e.g., DDT, aldrin, dieldrin, endrin, heptachlor, toxaphene); to permanent (e.g., lead, mercury, arsenic). Presumably, the less-persistent types should be more desirable, other things being equal; but those that degrade rapidly, such as organophosphate insecticides, are extremely toxic and nonselective, which encourages rapid emergence of resistant pests and destroys natural enemies. It is difficult to adopt chemicals that function without any drawback or disadvantage. As a result, pesticide use entails risks as well as benefits; if used improperly, pesticides can result in indoor environmental pollution.

No indoor air concentration standards for pesticides have been set. Pesticide products should be used according to instructions provided by the manufacturer. Pesticide labels indicate the potential harm to humans that could result with improper exposure. The CAUTION label indicates the least potential harm. The WARN-

ING label is next,with the DANGER label indicating that exposure to skin or eyes could cause severe burns. Pesticide labels also indicate the ingredients, detail the potential for environmental damage, and provide instructions for first aid, storage, and disposal. If chemical pesticides have been used in an area where an occupant has become sick, then the information should be provided to the physician who is handling the case.

Pesticides are applied in the form of sprays, powders, crystals, balls, and foggers. Insecticides and disinfectants are those used most often. Studies suggest that 80 to 90 percent of most exposures to pesticides occur indoors and that measurable levels of up to a dozen pesticides have been found in the air inside homes. The reason for this is that pesticides can contaminate indoor air from other sources, including contaminated soil or dust that floats indoors or is tracked in from the outside, pesticide containers stored indoors, and surfaces that collect and then release vapors from the pesticides.

Since most indoor pesticide exposure occurs via inhalation of spray mists, chemical pesticides should be used only outdoors or in well-ventilated areas. Follow instructions for recommended amounts of a pesticide. If preparation of pesticides requires mixing or dilution, this should occur outdoors or in a well-ventilated area. Pesticides should be stored in tightly capped containers. Contact local health departments to determine where to safely dispose of unwanted pesticides.

Most pesticides contain a high percentage of inert ingredients. These are used to make pesticide formulations less concentrated and more sprayable. EPA has identified more than 100 substances used as inert ingredients in pesticides as toxic. When pesticides are used in a situation that could impact the indoor environment, it is important to identify the inert ingredients.

The building IEQ management plan should characterize pest-control procedures by detailing frequencies, locations of applications, types of pesticides used, and the manner of application. The most effective control strategy for minimizing exposure to chemical pesticides is an approach known as integrated pest management (IPM). IPM uses a combination of pest prevention, chemical pesticides, and nonchemical pest controls to control pests with the least possible hazard to people, property, and the environment.

If hiring a pest-control company, evaluate the company's track record, insurance coverage, licenses, affiliation to professional pest-control associations, and proposed treatment methodologies. Identify the pest and use a pesticide targeted for that pest only in the amount directed, at the time and under the conditions specified, for the intended purpose. Whenever possible, use nonchemical methods of pest control. Ventilate the area during and well after pesticide use. Dispose of unused pesticides in a manner consistent with local and federal guidelines. Questions regarding pesticide use and safety may be referred to the National Pesticide Telecommunication Network (800-858-PEST).

The first step in IPM is to identify the pest problem. Pest specialists and library reference books can serve as information resources. The pest problem should be evaluated to determine if the pest should be eliminated or controlled.

Controlling a pest problem usually requires a less-intensive response than complete elimination. Control of pests can include operations as simple as removing standing water and pests' hiding places to reduce potential breeding sites.

If the scope of building management includes landscaping operations, then IPM can be used to minimize the use of chemical pesticides (i.e., source management). Native outdoor plants are good choices for landscaping because they tend to adapt well to local conditions with a minimum of care.

The EPA (US EPA, 1995) suggests six basic steps as part of a preventive health care program that will maintain a healthy lawn and reduce the need for chemical pesticide use:

1. Develop a healthy soil with the right pH, key nutrients, and good texture.
2. Choose a type of grass that grows well in your climate.
3. Mow high, often, and make sure the lawn mower blades are sharp. Lawns with longer grass (add about 1 inch) are healthier, resulting in fewer pest problems. When grass is between 22 and 32 inches, weeds have a hard time taking root.
4. Water deeply but not too often.
5. Prevent thatch buildup. Thatch buildup between grass blades and soil prevents water and nutrients from getting to the soil. Thatch can be raked out or broken up with a machine that pulls plugs out of the soil.
6. Set realistic weed and pest control goals. It is nearly impossible to get rid of all weeds and pests. Earthworms, spiders, millipedes, and various other beneficial microorganisms that help maintain a healthy yard can be destroyed though efforts to get rid of all weeds and pests.

In some cases, nonchemical control methods can work more effectively than chemical control methods. Your local county cooperative extension service, nursery, or garden association can provide information on biochemical pesticides and how to attract beneficial pest control "partners." For example, birds and bats can help control insects; ladybugs eat aphids, mealybugs, whiteflies, and mites. Other beneficial bugs include spiders, centipedes, ground beetles, lacewings, dragonflies, big-eyed bugs, and ants (US EPA, 1995).

Nitrogen Oxides

Nitrogen oxides are a group of gases produced by the combustion of fossil fuels, especially when combustion takes place at high temperatures and pressures. Combustion of petroleum products by motor vehicles and electricity-generating stations, paper mills, wood burning, waste incineration, oil refining and gas production, cement processing, steel and iron industries, and ethanol production are major sources. Indoors, primary sources are combustion processes, including unvented combustion appliances, vented appliances with defective installations, and tobacco smoke.

The two most prevalent oxides of nitrogen are nitrogen dioxide (NO_2) and nitric oxide (NO). Both gases are toxic, with NO_2 a highly reactive and corrosive oxidant. NO gradually reacts with the oxygen in the air to form NO_2. Direct emissions are mainly in the form of NO with smaller amounts of NO_2 and nitrous oxide (N_2O). Nitric oxide and nitrogen dioxide are frequently lumped together as NOx. NO_2, if abundant, can lend a brownish color to urban pollution. It is corrosive and a strong oxidizer. NO_2 may combine with water in the air to produce acid rain and, when combined with hydrocarbons in the presence of sunlight, can contribute to the formation of ground-level smog. NO persists for only a short time before reacting with ozone to form nitrogen dioxide. NO is colorless, odorless, and only slightly soluble in water.

NO_2 can irritate the lungs and lower resistance to respiratory infection, has an adverse effect on materials (e.g., causes corrosion of metals, fading of fabric dyes, degradation of rubber), and can damage vegetation. N_2O is not considered a directly harmful air pollutant in ambient air indoor.

There are no standards for nitrogen oxides in indoor air. ASHRAE and the U.S. National Ambient Air Quality Standard lists 100 micrograms per cubic meter ($\mu g/m^3$) (.053 parts per million [ppm]) as the average long-term (1-year) limit for NO_2 in outdoor air.

Venting the NO_2 source to the outdoors is the most practical measure for responding to existing conditions. Manufacturers are developing devices that generate lower NO_2 emissions. When equipment is scheduled for replacement, this should be factored into the selection of new equipment.

Carbon Monoxide

Carbon monoxide is a chemical compound containing carbon and oxygen with the formula CO. It is a colorless, tasteless, odorless gas. CO quickly enters the bloodstream, where it has an affinity for hemoglobin in the blood, and interferes with the delivery of oxygen to the organs and tissues of the body, resulting in asphyxiation. The health threat of exposure to high concentrations of CO is most serious for those who suffer from cardiovascular disease. Healthy individuals are also affected at higher concentrations. Exposure to high concentrations of CO is associated with impairment of visual perception, work capacity, learning ability, manual dexterity, and performance of complex tasks.

Carbon monoxide is formed whenever carbon or substances containing carbon are burned with an insufficient air supply. As little as 1/1000 of 1 percent (10 ppm) of carbon monoxide in air can produce symptoms of poisoning, and exposure to as little as 1/5 of 1 percent (200 ppm) could be fatal in less than 30 minutes. Significant sources include motor vehicles, industrial fuel combustion, home heating systems, and burning of solid waste. Motor vehicles are usually the major source, particularly auto, truck, or bus exhaust from attached garages, nearby roads, drive-through windows, or parking areas. High concentrations of carbon monoxide may be found in congested urban settings, especially in winter, since vehicle engines produce more pollution in cold weather. Smokers probably experience greater exposures. Away from large cities, CO tends not to be a problem pollutant in the outdoor environment.

Incomplete oxidation during combustion in gas ranges and unvented gas or kerosene heaters may cause high concentrations of CO in indoor air. Worn or poorly adjusted and maintained combustion devices (e.g., boilers, furnaces) or an improperly sized, blocked, disconnected, or leaking exhaust flue can result in significant CO exposures.

No standards for CO have been agreed upon for indoor air. The U.S. National Ambient Air Quality Standards for outdoor air are 9 ppm (40,000 milligrams per cubic meter [mg/m^3]) for 8 hours, and 35 ppm for one hour.

To control CO exposure, be sure combustion equipment is properly maintained and adjusted. Vehicular use should be carefully managed adjacent to buildings. Additional ventilation or recirculation of uncontaminated air can be used as a temporary measure when high levels of CO are expected for short periods.

Carbon Dioxide

Carbon dioxide (CO_2) is a colorless, odorless, and tasteless gas at normal temperatures and pressures. CO_2 is found in the ambient environment in ranges from about 300 to 400 ppm. The level of carbon dioxide is often used as an indicator for odor control in an indoor environment.

Combustion and human respiration are common sources of CO_2. CO_2 is present in all buildings. When CO_2 levels in an occupied building increase, it is usually because people are adding more CO_2 to the building environment than the HVAC system can handle (by dilution with outside air). ASHRAE Standard 62-1989 recommends 1,000 ppm as the upper limit in occupied buildings for comfort (odor) reasons.

Ventilation with outdoor air is used to control CO_2. It is important to note that CO_2 is used as an indicator and is not considered harmful at levels encountered in normal indoor environments.

ALLERGENS AND PATHOGENS

Biological material, including bacteria, viruses, fungi, mold spores, pollens, skin flakes, and insect parts, are ubiquitous in indoor environments. These particulates range from less than one to several microns in size. For comparison, a hair is about 30 to 70 microns in width. When airborne, biological material is usually attached to dust particulates of various sizes.

Biological hazards, or biohazards, consist of pathogenic microorganisms and similar substances that can pose a risk to the health and physical well-being of humans, animals, or other biological organisms. In the past, this definition was limited to infectious pathogenic microorganisms responsible for common communicable disease. The realm of biohazardous agents can now be expanded to include the following agents with the potential for causing disease: oncogenic viruses, recombinant DNA molecules, animals and plants and their by-products, and microorganisms of a rare or exotic nature (i.e., fungi, yeasts, algae). The presence of these biohazards in the environment is inevitable, even in the best-maintained buildings.

Fungi

Fungi are a ubiquitous, diverse group of organisms. Most do not cause disease (i.e., nonpathogenic) and are used either in their natural form or to produce other useful products: edible mushrooms, the antibiotic penicillin, yeast for making bread and beer, and Camembert and Roquefort cheeses. Fungi are made up of eukaryotic cells (complex cells with true nuclei) similar to those found in higher plants and animals. Fungi are like animals in that they are heterotrophic organisms that must consume organic matter to function. Fungi can consume dead organic matter or absorb tissue from living organisms.

Some fungi such as actinomyces and histoplasmosis, can cause mild disorders. *Stachybotrys* spp., *Aspergillus* spp., and *Cryptococcus* spp. have been identified as potential sources of indoor air quality problems. Some species of fungi are capable of producing mycotoxins, toxic metabolic by-products. These toxic by-products can occur in a variety of plant foods and in products derived from animals that have eaten contaminated feeds. *Aspergillus flavus* can release the mycotoxin known as aflatoxin, which is the most potent liver carcinogen known. Aflatoxin is a frequent contaminant of nuts, grains, and other crops and has been found in food intended for human consumption.

Bacteria

Bacteria are small, primitive, one-celled organisms called prokaryotes (cells whose genetic material is not enclosed by a nuclear membrane). Although very small (about 0.1 to 0.5 micrometers [μm]), most bacteria have distinctive cell shapes that affect their behavior and persistence. Some bacteria are beneficial (e.g., synthesize vitamins for the body, produce oxygen by photosynthesis), whereas others act as pathogens to humans or animals.

Bacteria have been the source of food and waterborne diseases (e.g., *Staphylococcus aureus, E. coli, C. botulinum, V. cholerae, Salmonella* spp.) and have been identified as a source of indoor environmental problems (e.g., *Legionella pneumophilia*).

Rickettsia

Rickettsial agents need living host cells in order to multiply. This host dependence causes them to be generally less hazardous than pathogens with less-stringent survival requirements. However, moderate-risk agents do exist; included in this classification are the rickettsias that cause such diseases as Rocky Mountain spotted fever and typhus. Chlamydiae act in a manner similar to rickettsias, causing infections such as psittacosis, a disease transmitted by birds that causes flulike symptoms in humans.

When birds use building air supply intakes as nesting sites, these intakes can be contaminated with bird droppings and other avian-associated problems, affecting the quality of the air brought into the ventilation system. For this reason, birds should not be allowed to roost near air supply intakes.

Viruses

Viruses are small (20 to 300 nanometers; a human hair is about 30,000 to 70,000 nanometers in diameter), simple genetic structures unable to change or replace their parts. Like rickettsias, viruses need living host cells in order to multiply. Viruses have been identified that can infect animals, plants, algae, fungi, protozoa, and bacteria. Viruses are classified on the basis of the hosts they infect. Because viral growth is tied to host cell functions, viruses are difficult to treat using medical therapy without causing some harm to the host cells as well. As a result, the relative risks of viruses vary more than those of any other biological agent.

Diseases caused by viral agents include influenza, measles, and mumps. Often, high incidences of flu or colds are perceived to be caused by poor IEQ, but these diseases tend to be transmitted from person to person and are probably not caused solely by the indoor environment. A medical evaluation of these issues is found in the next chapter of this book.

CONTROLLING BIOLOGICAL CONTAMINANTS

Any location that has moisture and nutrients can act as a reservoir for biological materials. Carpet, ceiling tiles, drapery, bedding, humidifiers, cooling coils, ductwork, and condensate drains can all act as biological reservoirs. Any place with standing water will contribute to the growth of bacteria and fungi. Condensation can provide enough moisture for growth. Cooling towers, water fountains, and jacuzzis can act as incubators of *Legionella* bacteria.

No microbial standards exist for general indoor air applications, except that ASHRAE recommends a relative humidity between 30 and 60 percent. Humidities below 45 percent will probably do a better job of preventing microbial growth. The American Conference of Governmental Industrial Hygienists (ACGIH) has published guidelines for the assessment of bioaerosols in the indoor environment and recommends comparing ambient microbial populations to those found inside during the investigation.

General good housekeeping and proper maintenance of heating and air-conditioning equipment are very important control measures, along with adequate ventilation and good air distribution. Higher-efficiency air filters can be effective in removing some allergens and particles from outdoor air. Integrated pest management and disinfectants should be used to control insect and animal allergens. Cooling tower treatment procedures exist to reduce levels of *Legionella* and other similar organisms. Maintaining indoor relative humidity below 60 percent can be helpful in controlling biological contaminants. Recently, the use of binary compounds for disinfecting areas with microbial contamination has met with some success.

Virtually everyone is exposed to biologic agents on a daily basis: Food, water, air, humans, insects, animals, and inanimate objects (e.g., sharp instruments) are a few common vectors capable of spreading pathogenic (disease-causing) organisms. In most cases, the human immune system is able to prevent the exposure from causing disease. Biological organisms, including fungal spores, bacteria, viruses, pollens, and protozoa derived from mold growth, have been identified in, and are also associated with, mammals, arthropods, and insects.

Moisture control is important. Exhaust fans in kitchens and bathrooms that are vented to the outdoors can eliminate much of the moisture that builds up from everyday activities. They can also reduce levels of organic pollutants that vaporize from hot water used in showers and dishwashers. Ventilating attics and crawl spaces prevents moisture buildup. Keeping humidity levels in these areas below 50 percent can prevent water condensation on building materials.

If using cool-mist or ultrasonic humidifiers, clean them according to manufacturers' instructions and refill with fresh water daily. Because these humidifiers can become breeding grounds for microbiological contaminants, they have the potential to cause diseases such as hypersensitivity pneumonitis and humidifier fever. Evaporation trays in air conditioners, dehumidifiers, and refrigerators should be cleaned frequently.

Water-damaged carpets and building materials can harbor mold and bacteria. Floods resulting in water-saturated carpets and building materials should be dried as soon as possible (within 24 hours). Even if the materials are dried and cleaned in a timely manner, removal and replacement should always be considered, especially in cases of repeated flooding. It is very difficult to completely rid such materials of biological contaminants once they are established. Good housekeeping practices can reduce but not eliminate dust mites, pollens, animal dander, and other allergy-causing agents. People who are allergic to these pollutants should use allergen-proof (easily cleaned) furniture whenever possible.

Radon

Radon is a colorless, odorless, radioactive gas. Radon decays into small radioactive particles that emit alpha radiation. These radioactive particles can be inhaled directly or attached to dust particles that are inhaled. Radon is measured in picocuries per liter (pCi/L).

Radon is a by-product of the radioactive decay of uranium. It is emitted from soil, rocks, and water. Since granites and shales tend to contain more uranium than other rocks, more radon comes from these types of rock. Building materials containing uranium can produce radon. The primary source of radon in buildings is the surrounding soil. Because radon is a gas, cracks, drain openings, and other penetrations can provide its entry into buildings. Radon is typically found in basements and lower-level crawl spaces.

Radon levels in outside air are about 0.3 pCi/L. Average residential levels are about 1 pCi/L. Levels in commercial buildings are even lower. EPA recommends taking action to mitigate radon if airborne levels exceed 4 pCi/L. ASHRAE Standard 62-1989 recommends levels not to exceed 2 pCi/L.

Building ventilation is a common strategy for controlling radon. Sealing foundations as a stand-alone strategy to prevent radon entry is rarely successful. However, sealing major entry points can improve the effectiveness of other methods of control. Increased outdoor air ventilation can reduce radon levels by dilution or by pressurization of the building. A ventilation-based strategy may not be very effective if the initial radon levels are greater than 10 pCi/L.

Lead

In its natural form, lead is a soft, gray metal that can cause severe health effects in relatively low concentrations in the body. Lead most noticeably affects the central nervous system, the kidneys, and the blood-producing organs.

Lead is very persistent in the environment, and exposure can come from food, water, contaminated soil and dust, and air. As a result of the addition of lead-based pigments to paint, a practice that ended in 1970, contaminated paint chips and dust are a common source of lead. Although adults can be affected, children are typically the population most at risk from exposure to lead. There are several reasons for this: Children's daily activities can involve contact with contaminated dust; a child's body retains a larger percentage of lead compared to an adult; and because a child is small, a particular amount of lead can have a bigger impact than the same amount would have in a larger person.

Lead is regulated in drinking water under the SDWA. Recent amendments have included some specific provisions for controlling lead in drinking water. One provision requires that only lead-free materials be used in new plumbing and in plumbing repairs; solders and flux are to contain less than 0.2 percent lead, and pipes and fittings are to contain less than 8 percent lead (these amendments are referred to as the Lead Ban). The federal Lead Contamination Control Act of 1988, P.L. 100-572 (LCCA of 1988), advised nurseries and child-care centers to test for lead content of drinking water consumed by children.

Although components used in a building can be considered "lead-free" in accordance with applicable regulations, newly installed brass plumbing components containing less than 8 percent lead as allowed by the LCCA of 1988 and the Lead Ban can still contribute to high lead levels in drinking water for a considerable time after installation.

The current EPA recommendation for lead in school drinking water is 20 micrograms per liter (μg/L) (EPA, 1991).

EPA recommends that action be taken if samples from drinking water outlets show lead levels higher than 20 ppb. Any drinking-water outlet with test results at or above this level should not be used for consumption until the source of contamination is found and corrective actions are taken to reduce the lead level below 20 ppb.

The Centers for Disease Control and Prevention has set 10 micrograms of lead per deciliter of blood (μg/dL) as the level of concern for lead poisoning in children. The Consumer Product Safety Commission (CPSC) has banned lead in paint. All other standards of lead exposure are for outdoor air, industrial workplaces, or occupational exposures. EPA is developing standards for abating existing lead-based paint.

Preventive measures should start with identification of lead-containing materials. An experienced professional should survey water, paint, and other suspected sources of lead. When lead has been identified, methods to reduce exposure include the following: removal of flaking paint by a reputable lead-abatement professional, maintaining intact paint by covering occasionally with lead-free surfaces, and flushing water that has stood in plumbing overnight for at least three minutes before the first draw of water in the morning. The following actions can help to reduce lead exposure: cleaning play areas, frequently mopping floors and wiping window ledges and other areas with damp cloths, keeping children away from areas where paint is chipped or peeling, preventing children from chewing on windowsills and other painted areas, and ensuring that hands are washed before mealtimes.

Dust

Dust is composed of particles in the air that settle on surfaces. Large particles settle quickly on skin and can be trapped by the body's defense mechanisms. Small particles are more likely to be airborne and are capable of passing through the body's defenses and entering the lungs.

Dust can be generated from multiple sources, including soil; fleecy surfaces; pollen; lead-based paint; and burning wood, oil, and coal. A major source of indoor dust may be remnants of human skin that is continually shed as the body constantly replenishes itself. Particles in the air may be of natural or industrial origin. Airborne particles are also sometimes referred to as *particulates* or *aerosols*. With regard to air-pollution measurements, these terms all mean essentially the same thing. There are many sources of particles. Combustion sources that result in the release of particles into the air include refuse burning; forest fires; industrial fuel use in boilers, kilns, pulp mills, refineries, and power-generating stations; and motor vehicles. Natural or nonindustrial sources include windblown soil dust, sea salt, dust from volcanoes, and fine sand from roads (especially when pulverized by traffic).

The results of health studies conducted over the past decade have dramatically increased the focus on particulate air pollution, in particular the importance of smaller-sized particles. The smallest particles have the potential to cause the most damage to human health because they penetrate the lungs, whereas larger sizes are filtered out by the nose and throat. Particles small enough to penetrate the lungs are termed *inhalable* particles. Particulate matter (PM) of this type composed of particles 10 microns or less in diameter are referred to as *PM10*. Particles classified as PM10 are less than a third the width of a human hair. The EPA National Ambient (outdoor) Air Quality Standard (NAAQS) for particles less than 10 microns in diameter is 50 $\mu g/m^3$ for an annual average and 150 $\mu g/m^3$ for a 24-hour average.

In 1997 rule changes, EPA set new annual and 24-hour NAAQS standards for particles classified as PM2.5 (i.e., particles less than 2.5 microns in diameter), retained the current annual PM10 primary standard, and revised the form of the current 24-hour PM10 primary standard. EPA stated that it based these changes on health information extracted from the agency's criteria document for PM. The document concludes that the evidence for PM-related effects from epidemiological studies is fairly strong, with most studies showing increases in mortality, hospital admissions, and respiratory-symptom and pulmonary-function decrements associated with several PM indices. While the results of the epidemiology studies should be interpreted cautiously, they nonetheless provide ample reason to be concerned that detectable health effects are attributable to PM at levels below the current NAAQS. EPA is proposing to establish two new primary PM2.5 standards—an annual standard set at 15 $\mu g/m^3$ and a 24-hour average limit of 50 $\mu g/m^3$.

It is significant that PM10 is the only major air pollutant for which standards are set based not on the chemical composition, but on the physical characteristics of the pollutant.

Use good housekeeping practices to keep dust to a minimum. Damp dusting and high-efficiency vacuum cleaners can be used for dust control. Vacuum cleaners with regular filters expel dust into the air. Since this dust could remain airborne for several hours after vacuuming, in areas where individuals have allergic reactions, make sure vacuuming is done shortly after the close of business to enable suspended dust particles to settle before the beginning of the next workday.

Upgrading filters in ventilation systems to higher efficiencies will provide cleaner air to the interior of a building. Typical filters remove 20 to 30 percent of airborne contaminants from supplied air. ASHRAE Standard 1989–62 recommends the use of filters that are 60 percent efficient. Before upgrading filters, ventilation systems should be evaluated to ensure efficiency of the overall system will not be affected. Filters should be checked and changed frequently. Relative humidities below 45 percent will help control dust mites and resulting biological dust. Combustion appliances should be exhausted to the outside.

Dust Mites

Many people with allergies are allergic to dust mites. Dust mites are found in bedding materials, cloth upholstery, and dust. They can be identified via air sampling or analyzing dust samples. Although microscopic analysis is possible, allergen assays are considered a better means of quantification.

Frequent cleaning and using easily cleanable furniture will reduce dust mite loads, as will removing carpet and fibrous furniture coverings. Acaricides can reduce dust mite populations, but this method of control could also result in chemical exposure. Maintaining humidities below 45 percent will also reduce dust mite populations.

SAMPLING FOR INDUSTRIAL HYGIENE

Industrial hygiene sampling can be used to determine levels of contaminants in buildings. Sampling should not be conducted beyond basic parameters (temperature, relative humidity, carbon dioxide, carbon monoxide, and particulates) unless there is a specific reason for monitoring, such as when a physician recommends identifying specific contaminants as a result of a medical evaluation. The next chapter addresses medical evaluations.

Two general categories of industrial-hygiene testing are performed to detect chemical air contaminants: direct reading measurements, which provide immediate results, and samples collected on-site and later analyzed in a laboratory. Laboratory measurements can take up to two weeks (or more) for results to be available. Immediate results enable the investigation to proceed at a more rapid pace, but laboratory tests are usually more reliable and enable identification of lower levels of contamination. When contracting consultants to conduct sampling, make sure a timeline is provided for obtaining results and that this information is made available to concerned occupants.

Often, colorimetric (detector) tubes are used to provide on-site direct-reading measurements for specific chemical contaminants. Suction devices (i.e., bellows or piston pump) are used to pull air through detector tubes filled with reagents that react to specific chemicals in the air. Air-concentration levels are indicated by the color change in the detector tube. Typically, the longer or darker the color stain, the higher the concentration. The chief advantage of this type of measurement is that it provides a quick and inexpensive means of obtaining information. It is called a grab sample because it measures contaminant levels at only one specific place and point in time. Margins of error tend to be about 20 to 30 percent. Detector tubes can be used easily by in-house personnel and measure carbon dioxide, carbon monoxide, formaldehyde, and various petroleum hydrocarbons. The bellows or piston pump usually costs between $300 and $400 and can be reused. Tubes are good for one test and cost about $5 each.

Smoke tubes provide a very cost-effective method of evaluating pollutant pathways and airflow. These tubes are connected to a squeeze bulb that is used to blow a white chemical smoke into the air. The direction of smoke movement indicates airflow and the directions a pollutant could travel as well as driving forces. Smoke tubes are supplied by the same companies that supply detector tubes. Sometimes people are sensitive to the chemicals used in the smoke tube test; this should be taken into consideration when scheduling the evaluation.

Electronic direct-reading instruments can be used to measure many indoor air contaminants (e.g., carbon dioxide, carbon monoxide, temperature, relative humidity, VOCs). Depending on the instrument, costs range from $5,000 to $15,000. Typical rental charges range from $50 to $150 per day. Often a consultant will rent this equipment and charge the client the rental cost plus a handling fee. Usually, these instruments can be connected to a continuous data recorder that provides measurements over time. This type of monitoring is effective because it provides a story of what is happening in a building, whereas grab sampling provides just a snapshot and may miss the event or events that are causing the problem. As a result, continuous monitoring, though more expensive, tends to be more effective than grab sampling.

Air-sampling pumps or passive dosimeters can be used to collect samples that are sent to laboratories for analysis. Typically, an industrial hygienist conducts sampling using pumps. Passive dosimeters do not use air-collection devices and are very easy to use. They are removed from a package and affixed to a person or located in a specific location for a specific period of time. Afterward, the device is delivered to a laboratory for analysis. Sometimes this method of testing is very cost-efficient. A drawback is that there can be a considerable time lapse before results are made available. It is also important that an environmental professional evaluate the testing results.

Often biological contamination is a potential culprit. There are no regulatory limits for exposure to biological contaminants, so it is difficult to make recommendations based on sampling results. Air sampling for bioaerosols offers only a snapshot and is difficult to interpret or quantify. Sometimes identification of allergens will help an investigation if someone is sensitive to a specific allergen. Bulk and swab sampling can help in identification of bioaerosols. ACGIH and the American Industrial Hygiene Association (AIHA) have published guidelines on assessing bioaerosols in the indoor environment.

If visible microbial contamination is discovered, clean and disinfect the contaminated areas. If cleaning is not effective, then the contaminated material should be removed. Microbial organisms are ubiquitous in the environment, so if evidence of contamination is visible, (e.g., mold or mildew), it should not be necessary to collect samples.

If medical evaluation indicates you are dealing with a microbial contamination problem, it is important to enlist the aid of someone who is knowledgeable in the field of microbiology in the course of the investigation. When microbial sampling is determined to be necessary, ensure that the firm that performs the sampling and follow-up has specific skills in the area of microbiological evaluation.

Simple, inexpensive, easy-to-use lead-detection tests (i.e., chemical spot tests) are available in the marketplace but should be used only by someone experienced in conducting lead surveys. Test surfaces must be scraped down past any layers of paint. Laboratory analysis of paint chips and portable detectors (X-ray fluorescence) are two other generally accepted methods of testing for lead in paint. Chemical spot tests and portable detectors provide on-the-spot results. Portable detectors are very expensive and can be used only by a specially trained person, but the testing does not affect the test surface. The paint-chip analysis must be performed at a laboratory and is the most reliable test. If in-house expertise is not available, a reputable consultant should be engaged to conduct the survey and make recommendations for proper removal if lead is identified. If your facility is covered by OSHA regulations, industrial-hygiene exposure monitoring must be performed to determine lead-exposure levels. If lead-exposure levels are above OSHA limits, workers must be protected. Even if your facility is not covered by OSHA regulations, employees and occupants should be protected from being exposed to adverse lead levels.

When asbestos is thought to be a problem, two types of samples can be used for identification: *bulk samples* and *air samples*. Bulk samples of material are collected to determine if asbestos-containing material is present. The samples need to be sent to a laboratory for analysis using polarized light microscopy. If asbestos has been identified or is suspected to be in an area, then air samples can be collected to determine airborne concentration levels. Asbestos air samples are analyzed using phase contrast microscopy and are usually sent to a laboratory unless someone has brought a microscope on-site for that purpose. It is important to remember that this method of analysis actually counts any type of fiber that is at least three times longer than it is wide and is greater than five microns in size. Fiberglass and other similar materials could be evaluated using this method. Another important consideration is to evaluate levels of detection or sensitivity. Inform whoever is collecting an airborne sample for analysis that a level of detection or sensitivity should be at least equal to expected background levels.

Transmission electron microscopy (TEM), a more expensive analysis that is specific for identifying asbestos, can be performed when contamination is suspected from other sources.

Radon levels can be tested using several different methods. These methods involve leaving a detector in place for anywhere from a few hours to several days. The detector is then analyzed in a laboratory. Costs can range from $25 to $150. Radon should not be a major concern unless you have occupied areas that are below ground level.

NOTES

American Society of Heating, Refrigerating, and Air-Conditioning Engineers (ASHRAE). 1989. ASHRAE Standard 62–1989, Ventilation for Acceptable Indoor Air Quality.

ASHRAE. 1993. *1993 ASHRAE Handbook: Fundamentals, I-P Edition.* Atlanta: ASHRAE.

Guidelines for Assessment of Bioaerosols in the Indoor Environment. 1989. Cincinnati: American Conference of Governmental Hygienists.

Illuminating Engineering Society (IES) of North America. *IES Handbook Reference Volume.* New York: IES

United States Environmental Protection Agency (US EPA). 1990. *EPA Fact Sheet: Ventilation and Air Quality in Offices—1990.* Washington, D.C.: U.S. EPA.

US EPA. January 1991. *Lead in School Drinking Water.* EPA 570/9-89-001.

US EPA. 1995. *Citizen's Guide to Pest Control and Pesticide Safety.* EPA-1995.

US EPA and United States Consumer Product Safety Commission Office of Radiation and Indoor Air (6604J). April 1995. *The Inside Story: A Guide to Indoor Air Quality.* EPA Document #402-K-93-007.

US EPA and National Institute of Occupational Safety and Health. December 1991. *A Guide for Building Owners and Facility Mangers.* EPA/400/1-91 033.

US EPA, Office of Acid Deposition, Environmental Monitoring and Quality Assurance. 1987. *Project Summary: The Total Exposure Assessment Methodology (TEAM) Study.* EPA-600-S6-87-002.

Chapter 3

Investigating Health Complaints

Ronald E. Gots, M.D., Ph.D.

Purpose

Though many books, chapters, and articles have been written about indoor air and health, they all lack a practical how-to quality. The purpose of this chapter is to provide a building manager with useful information to help assess the severity of complaints and devise an effective response strategy. To do so, those responsible for maintaining office spaces must understand some of the medical principles that underlie indoor air issues; these will be presented simply and practically. The ultimate purpose of this chapter is to keep those with financial responsibility for a building from doing too little or too much. Either can lead to disastrous financial consequences.

Introduction

Indoor air issues begin with health or comfort complaints. Someone in the office (or perhaps many people) complain to the office manager, supervisor, or building manager. The complaints may involve pure comfort allegations—too hot, too cold, too dry—or they may involve more-specific complaints—headaches, burning eyes and nose, red eyes, cough, fatigue, nosebleeds. On rare occasions complaints may be even more dramatic—for example, mass faintings. Since these complaints are first fielded by a medical layperson, that person has the unenviable responsibility of becoming a de facto medical triage officer. The initial task is to decide whether these complaints indicate a serious medical problem or whether they are readily managed by a nonmedical person.

Even more vexing is the dual nature of such complaints. On the one hand, complaints may be due to actual physical problems associated with the environment. Alternatively, they may be due to various psychological and/or emotional factors. Or they may be a combination of the two. The latter is no less concerning than the former and in fact is probably both more common and more serious to the integrity and financial risk of the organization and the building owners. Table 3–1 presents many of the potential causes of symptoms in building occupants. Note that symptoms people associate with the building may actually be related to the building, or they may not. Though people perceive that a building is associated with symptoms, those perceptions have varying degrees of accuracy. A good rule of thumb is that as time goes on, the accuracy of those perceptions diminishes. In other words, once a crisis has arisen and concerns are widespread, people will associate more and more symptoms with the building, whether or not those symptoms are actually related.

Secondly, Table 3–1 indicates that even symptoms that are accurately associated with the building may or may not have anything to do with indoor air quality (IAQ). The rush to judgment that "the air is bad" has only a random

Table 3–1.   Causes of symptoms in building occupants: IAQ related, IAQ unrelated, and building unrelated.

IAQ	Non-IAQ
Building related	
Stagnant air	Thermal
Humidity	Lighting/Noise/Ergonomics
Odor	Psychosocial
Irritant	Political
Allergen	
Pathogen	
Building unrelated	
Environmental and home allergens	
Medications	
Underlying disorders	
Unrelated events (i.e., cancer and miscarriages)	

chance of being correct, without a complete investigation of the medical issues and proper correlation with environmental findings (Baker, 1989; Gots, 1993; Lees-Haley, 1993).

THE MANY CAUSES OF SYMPTOMS: DIFFERENTIAL DIAGNOSIS

The essence of medical practice is the differential diagnosis. By this we mean that a constellation of symptoms leads to a number of considerations about possible causes. This in turn leads to a series of diagnostic tests to rule in or rule out any of the potential causes.

For example, if you complain to a physician about a headache, the physician will get a more detailed history from you and arrive at a preliminary differential diagnosis. That differential diagnosis may include a brain tumor, eyestrain, a cervical strain, a migraine, a sinus infection, stress, or many other conditions, any of which can cause headaches. Appropriate tests follow to rule out the most serious causes, such as a tumor. The process of evaluating workers with complaints is no different. Each symptom has many possible causes that can be ruled in or out only through a careful history, physical examination, and proper testing targeted to the differential diagnosis.

Unfortunately, indoor air complaints are only rarely evaluated in this fashion. Frequently, the first person involved is a heating, ventilation, and air-conditioning (HVAC) engineer, a maintenance person, or an environmental consultant. Thus, the decision that the problem is related to air is often made at the moment the complaint is initiated. That is a bit like sending everyone with a headache to a neurosurgeon to explore the brain for a tumor. Since many worker complaints have nothing to do with indoor air, many of these investigations assume, incorrectly, that poor air quality is responsible. Although building management must recognize this potential for error, cost and practicality demand that the simplest, most cost-effective approach be followed. This means that every symptom an office worker reports cannot support a full medical evaluation. Therefore, it is incumbent upon consultants and building engineers to know when to bring in medical help and when not to.

Symptoms in workers are often called *health effects*. This term is inappropriate when introduced too early, because it makes the unsupported assumption that a symptom is the "effect" of something in the environment, when that remains to be proven. It also assumes that every complaint has something to do with health. Discomfort is not the same as ill health. A person may find a room too cold or too hot, hence uncomfortable. Or someone may have a minor symptom such as a transient headache or fatigue. Absent an underlying physical disorder, none of these situations can be said to imply an adverse health effect. Symptoms that workers may associate with the workplace are often quite varied in nature, having little to do with one another or with a common cause. Figure 3.1 illustrates the chaotic and diverse nature of symptoms or disorders that may be reported by office workers and that workers may relate to the workplace.

In the case of symptom complaints associated with office buildings, the differential diagnosis is complicated by a multidimensional consideration: the need to determine a diagnosis for the worker as well as for the building's condition and ultimately, to integrate the two. Not only are we trying to learn whether the headache is due to eyestrain or a brain tumor, we are also trying to determine whether environmental and/or psychosocial factors are contributory. Such a differential diagnosis is also complicated by multidisciplinary considerations. The person who diagnoses causes of headaches is not generally the one who decides whether the HVAC system is working properly, nor should he be.

Matching the Symptoms with the Possible Causes

Most health complaints begin with one or more workers who decide that the office is creating health problems. At the outset, they have made their own diagnosis and have determined the cause. As often as not, this attribution is incorrect, and it is important for the investigator to understand this.

However, it is equally important to realize that once a belief is firmly in place, it may be hard to dislodge, and indelicate attempts to do so may create resentment and distrust. In other words, you had better have good data as well as a caring manner when discussing with workers potential causes for their problems, which may differ from what they have come to believe.

Figure 3.1. Symptoms and disorders commonly related to the office environment by workers: A chaotic mixture

Headache

Rash

Itchy eyes and throat

Nosebleeds

Fatigue

Miscarriages

Asthma

Fainting

Cancer

Trouble concentrating

In other situations, there may be many causal attributions—each affected worker may have a unique explanation as to what is responsible for his or her symptoms. Quite often, these causal attributions change with time—either the workers develop new ideas or data provide leads to new understandings, or both.

Often, the first management person notified accepts uncritically the proposition that the symptoms and disorders are related to the indoor air, if that is what the concerned workers believe. While it may be appropriate to respond that way—to accept attributions from workers and occupants—it is equally important to keep an open mind and be aware of the many possible causes. In general, the broader the range of manifestations or symptoms, the less likely it is that the building is responsible for all of them. Among the complaints may be some that *are* connected to the environment. But others may be unrelated, only thought to be associated with the workplace. Lumping varied symptoms together as "health effects" and attributing them to bad air can be overly simplistic and fraught with error. Table 3–2 demonstrates this point by delineating some of the diagnostic considerations that could be connected to a given symptom absent any additional information.

This table illustrates several things. First, many of these symptoms and disorders do not share common possible causes. Second, in some instances the indoor environment has no known connection to the disorders. Third, all of the disorders and symptoms have multiple possible causes, many of which are unrelated to the office. For those that may be office-related, many are unrelated to IAQ. The important message here is that symptoms and disorders are not necessarily IAQ health effects just because someone has decided that they are. A thorough exploration of cause requires a differential diagnosis, a physical examination, and appropriate testing of the patient before even a possible link can be made. After that is completed and potential environmental causes are identified, environmental testing may or may not find the factors that could explain an individual's symptoms. Although the best explanation for symptoms requires a comprehensive set of evaluations, that is neither practical nor necessary in every instance. We will see in subsequent sections how to triage and how to limit and focus investigations (Abend, 1995; International Society of Indoor Air Quality and Climate [ISIAQ], 1996; United States Environmental Protection Agency [USEPA], National Institute of Occupational Safety and Health [NIOSH], 1991).

Effective Triage

Symptom complaints are often minor and may respond to simple adjustments in airflow from the HVAC system or small changes in temperature and/or humidity. These are obviously simpler solutions than bringing medical and engineering

Table 3–2.   Possible causes of symptoms and disorders often attributed to IAQ.

Symptom or disorder	Common causes	Office-related possible causes	IAQ possible causes
Headache	• Stress • Eyestrain • Sinusitis • Migraine • Neck strain	• Stress • Eyestrain • Psychosocial	• Rarely chemicals
Rash	• Insect bite • Eczema • Contact dermatitis • Other skin disorders	• Neurodermatitis (stress-related)	• Fiberglass
Itchy eyes	• Contact lens problems • Allergies • Infection	• Eyestrain	• Low humidity • Mold • Chemicals • Dust • Fiberglass
Nosebleeds	• Allergies • Infections • Trauma		• Low humidity
Fatigue	• Many serious diseases • Depression • Sleep deprivation • Chronic fatigue syndrome	• Boredom • Job dissatisfaction • Overwork	• Possible (rarely volatile chemicals)

Miscarriages	• Idiopathic
	• Various factors
	  Genetic
	  Structural
	  Infection
	  Metabolic
	  Etc.
		• None known	• None known

(continued)

Table 3–2. Continued

Symptom or disorder	Common causes	Office-related possible causes	IAQ possible causes

Asthma	• Allergies
	  Cat
	  Dog
	  Dust at home
	  Pollens
	  Etc.
	• Exercise induced
	• Cold air
			• Allergies
			  Dust
			  Mold
			• Rarely irritant chemicals

Cancer	• Smoking • Heredity	• None known	• None known
Trouble concentrating	• Many serious diseases • Depression • Sleep deprivation • Chronic fatigue syndrome	• Boredom • Job dissatisfaction • Overwork	• Possibly (rarely) volatile chemicals

Fainting	• Blood
	  Pressure
	  Abnormalities
	• Heart disease
	• Anxiety
		• Anxiety	• Major chemical intoxication (i.e., carbon monoxide)

consultants into every office in which workers have complaints. The challenge is for the first consultant on the scene (often the building maintenance staff) to do an effective triage or initial assessment: that is, to determine when a problem is trivial or serious and to recognize and respond quickly to any escalation. The key qualities needed in such a person to serve these functions well are common sense, an understanding manner, and sufficient awareness of the possible underlying causes of complaints. Independent assessments by outside consultants are generally not required in the majority of complaint situations. Nevertheless, consultants can serve as a sounding board or provide independent confirmation of an in-house assessment. Should a situation deteriorate, however, ready access to the appropriate consultants can be critical.

Because situations are so varied, there are no absolute triage rules that can be applied universally. However, several important clues and rules of thumb can help guide appropriate action.

Triage considerations are twofold. The first aspect might be called the *who* question. Who should do the investigation and work at resolving the problem? The second is the *what* question. What needs to be done to address concerns and resolve the problem?

One rule is that the development of indoor environmental complaints over time is unpredictable. What starts as a seemingly minor problem can become a nightmare, or what begins with explosive outbursts can fizzle. Second, building managers and owners can get into trouble by either doing too little or doing too much. Failing to recognize the potential severity of a problem, minimizing people's concerns, and responding too tentatively may increase anger, distress, symptoms, and financial risk. Conversely, overreacting at the first notice of a complaint can produce data that have no meaning or unleash frightening hazardous materials (HAZMAT) responses, both of which lead to heightened anxiety, expensive remediation, and psychologically induced illnesses.

Because of the risks just described, one thing should be clear. Whoever is entrusted with the preliminary evaluation of these complaints needs to understand the risks, have good common sense, and be an effective communicator. The management of perceived indoor air issues is half a technical function and half public relations. If these traits can be found in someone in the building-maintenance department, then that individual may be the appropriate initial contact. If on the other hand the building-maintenance department has no such individual, then the owners and managers would be well served to involve a consultant immediately. In either case, the need to have an IAQ plan in place *before* problems arise cannot be overstated.

Decisions about how to proceed with an investigation and what level of investigation to undertake are determined by a number of factors. The levels of investigation will be discussed in the next section. The relevant factors include the severity and nature of symptoms, lost work time, attributions made by employees, the number of people complaining, how long the problem has been going on, what has been done to address the problems thus far, whether consultants have been involved, and the quality of the landlord-tenant relationship.

Severity and Nature of Symptoms

The very first question that must be asked and answered is, "How sick are people?" Obviously, there is a difference between loss of consciousness or hospitalization for Legionnaires' disease and complaints of headaches. The first two may require evacuation of the building and primary attention to health needs. The second generally permits a more measured and systematic response. Severity of health complaints is, therefore, an initial triage question. If complaints consist of only minor symptoms, it may be possible to manage the problem locally and without medical intervention. If symptoms are more dramatic, and/or if workers have gone to hospitals or have seen physicians about them, the building manager must include appropriate medical help in the consulting team (ISIAQ, 1996; Hodgson, 1995).

This triage consideration is clear: severe symptoms demand appropriate referral. Ironically, however, the expected corollary—severe symptoms equal a serious environmental problem—is less often true. There is far less correlation than one might think between apparent severity of symptoms and long-term health allegations, litigation, and financial risk to the building owners. Numerous cases of mass faintings with emergency responses by HAZMAT, fire departments, and health departments have uncovered no evidence of environmental hazards (Bauer et al., 1992; Spitters et al., 1996). In many such cases, diagnoses of "mass psychogenic illness" have been made with minimal long-term cost to the facility (Alexander and Fedoruk, 1986; Brodsky, 1983; Hall and Johnson, 1989; Light and Tiffany, 1991).

On the other hand, many matters have produced catastrophic losses to building owners, architects, facility managers, ventilation and design engineers, and contractors that were heralded at first by seemingly trivial health complaints. In a matter that I investigated, several workers left a building complaining of headaches following interior renovation. Two years later, one of those workers settled a lawsuit for $400,000. Therefore, although one can offer triage advice, no rules can eliminate economic risks, even when health risks are not significant.

Lost Work Time

Lost work time is an important clue to the severity of the problem. If people believe that they are too ill to work, or if they are afraid to return to the building because they think it makes them sick, you are confronting a serious problem. You should also note if workers are restructuring their workday (e.g. leaving early) or restructuring their workstations to avoid what they perceive to be the problem. Since "the problem" may include actual health disorders and/or liability for lost productivity, lost work time and refusal to enter the building are measures of severity. One of the first questions to ask, therefore, is "Are people staying out of work?"

Attributions Made by Employees

If early complaints suggest that either the workers have *not* attributed symptoms to environmental factors or they have suggested an easily correctable change (e.g., too hot, too cold, or insufficient humidity), the situation may be manageable at the facility level.

On the other hand, if workers are using terms such as "sick building," "poisoning," or "toxic," this implies a potentially more difficult-to-manage situation. The more dangerous the building is in the eyes of the workers, the more complicated the resolution of the problem. The triage question is then "What do people think is wrong?"—a question that should also be asked of workers on an individual basis.

Number of People Complaining

Complaints from only one or two people may be more easily managed than a companywide problem. It is important to remember, however, that symptoms may herald an early health problem that can become more widespread and that symptoms can be contagious through the psychology of suggestion; one or two complaining workers can quickly spread their symptoms to others. Thus, it is important to respond in a caring and competent manner even when there is just a single complaint. The most intense response will be stimulated by situations with the most complainants.

Duration of Problem/Involvement of Other Consultants

A critical rule of thumb is that the longer the problem has existed, the more resistant it will be to correction. In addition, sometimes the building owner/manager is the last one called. The tenants have brought in their own consultants who have been unsuccessful in resolving the problem. Problems that are firmly entrenched generally require sophisticated and experienced problem solvers.

Quality of the Tenant-Management Relationship

Indoor-air allegations are a growing source of leverage in landlord-tenant disputes. It is not uncommon for leases to be broken based upon "bad air," and tenants have brought major lawsuits against landlords for such problems. In some cases tenants have trumped up "bad air" allegations in order to terminate a lease. I have personally investigated at least one such matter. Recognizing this, if you are a building owner or manager, the quality of your relationship with your tenant becomes an important triage issue. If the relationship is bad, or if indoor-air complaints follow a series of other problems, consider early expert consultation rather than simple fixes.

INVESTIGATION

The bottom line of any investigation is to make the workers feel more comfortable—to take their symptoms away. Sometimes this can be done without a complete understanding of the reasons that they have symptoms; a minor adjustment may satisfy people. At other times, the investigation of health complaints and their causes requires a systematic multistage process. In these instances, symptoms must be evaluated, the environment must be evaluated, and the two must be correlated as accurately as possible. While it is important to realize that an engineer cannot definitively determine the cause of someone's headache or other symptoms, some worker concerns may be appropriately addressed and solved by an engineer. Thus, if a worker feels that the air is too dry, a limited and focused evaluation of building humidity and correction where indicated is the most cost-effective and logical approach.

To simplify our organizational discussion, we may consider three levels of investigation. The level required will be dictated by the triage considerations identified in the preceding section.

Level 1

At its simplest, an indoor-air investigation involves an uncomplicated inspection and minimal corrections. When triage suggests a minor problem, then that is what should be addressed. It may be accomplished without sophisticated consultations or medical input. It may involve implementing an operations and maintenance plan; minor cleaning of the HVAC system; adjustments to airflow, temperature, or humidity; or any combination of these measures. For example, workers may express concern about localized mold growth. Cleaning the mold and adjusting the humidity may suffice.

Level 2

The second-level investigation requires more intensive analysis. Here there may be more health complaints, and triage may suggest a more serious problem. At this level appropriate consultants need to be engaged. These consultants must have the skills, knowledge, and expertise to solve complex problems involving engineering, industrial hygiene, and medicine—the three disciplines that are fundamental to office environments and health. At Level 2, a more intensive search for causes is in order and should include medical interviews, evaluations of occupational stressors, and facilities evaluation. Sampling for airborne contaminants is generally not performed unless indicated by point source evaluations; rather, general HVAC issues are evaluated and potential point sources of contaminants are reviewed (Persily, 1994; ISIAQ, 1996; Ventresca, 1995).

Level 3

The most intensive evaluation occurs at Level 3, where all of the expertise and evaluations noted previously are performed, but comprehensive environmental sampling and laboratory analysis may be required. A summary of the critical elements of this level of investigation are shown in Table 3–3.

A comprehensive discussion of these phases is beyond the scope of this chapter; some general descriptions are provided.

Complaint Evaluation

Complaint evaluation is the process of cataloguing the worker's symptoms and their attributed causes. But it also includes (as part of the differential diagnostic process) asking about other factors, including home environment, prior allergies, and job satisfaction. In general, widely distributed questionnaires, while commonly used, are not a good idea. Unless they are extremely well constructed (few are) and properly administered, they can provide leading questions and can make any building look sick (Gots, Gots, and Spencer, 1992; ISIAQ, 1996; Quinlan et al., 1989; Samimi, 1995). Brief interviews are strongly recommended. After reviewing complaint records, interview a representative number of complainants and noncomplainants. Additionally, occupants (complainants and noncomplainants) may be asked to maintain a diary of environmental conditions and their personal concerns. Complainants tend to keep more detailed diaries, and although this would be considered a bias in a formal scientific study, in a com-

---

Table 3–3.   Phases of a comprehensive (Level 3) investigation.

1. Complaint evaluation

2. Clinical evaluation

3. Source evaluation

4. HVAC evaluation

5. Sampling (if necessary)

6. Causation analysis

7. Communication (this applies to levels 1 and 2 as well)

plaint investigation the diaries can be correlated with a daily log of building conditions. This may lead to further insight into occupant concerns. Having occupants maintain diaries also involves them in the investigation process.

Clinical Evaluation

Clinical evaluation refers to the medical examination of the workers. At the simplest level this may mean looking at a red throat or a skin rash. At its more complex, it may include ordering blood or skin tests to evaluate allergies, chest X-rays, and pulmonary function tests.

Source Evaluation

Source evaluation refers to the process of examining potential sources of emissions or contamination. At times, the workers themselves will point to a perceived source—for example, recent painting or other renovation, copy-machine chemicals, new carpeting, and mold.

HVAC Evaluation

HVAC evaluation includes examination for dirt, dust, and biological contamination (e.g., mold growth); evaluation of the registers—location, cleanliness, balance, and so on; determination of the quality of airflow at specific locations; determination of the mix of outdoor and indoor air; and temperature and humidity evaluation (American Society of Heating, Refrigerating, and Air Conditioning Engineers [ASHRAE], 1989; ASHRAE, 1981; ISIAQ, 1996; Persily, 1994; Sheet Metal and Air Conditioning Contractors' National Association [SMACNA], 1995; Ventresca, 1995).

Sampling

Sampling follows the previously discussed evaluations. It must be targeted and specific and should be done only if two conditions are met: contaminants are suspected, and those contaminants would likely explain the symptoms. Random, extensive sampling should never be permitted. For example, sampling and analysis using gas chromatography (GC) and mass spectrophotometry (mass spec) can yield a list of 300 volatile organic chemicals at low parts per billion levels that have no medical relevance. This will inevitably increase the cost of evaluation and potentially create unnecessary distress.

Causation Analysis

Causation analysis in this case refers to the process of putting environmental data together with clinical data to reach cause-and-effect conclusions. Often this attempt is made by individuals with no relevant medical expertise who then draw incorrect conclusions. For example, an HVAC engineer may find poor airflow/air distribution in an area and conclude that it caused the reported symptoms. If symptoms include skin rashes, this conclusion is wrong, because airflow problems do not cause skin rashes. Or an industrial hygienist may find very low levels of a variety of volatile organic chemicals and conclude that they caused headaches, when they could not have done so. Or a small amount of visible mold may be blamed for respiratory complaints despite the fact that the sufferers were not allergic to that mold.

This process of causation analysis is the most sophisticated part of the investigation, for it goes beyond data collection and into differential diagnosis and clinical interpretation. Many engineers and industrial hygienists do this very poorly, but so too do many physicians who are unfamiliar with the health issues associated with office buildings. That is why a multidisciplinary team approach that includes consultants with indoor air expertise is so important.

Communication

Communication is listed last, but it is not the least of the phases. Rather, effective communication must be a continuous process. From the start, workers need to understand the investigation process itself. Throughout every phase of the procedure, the workers need to know that explanations are being sought and relief is on its way. This aspect of the investigative activity is extremely important, for it may determine how smoothly and inexpensively the situation is resolved. Occupants who are involved in and informed about the investigation are more willing to accept and more likely to understand investigative conclusions and recommendations. Disgruntled workers who feel sick and worried and who do not believe they are being heard or understood can quickly become a large group of plaintiffs.

Effective communicators need to be good listeners, but they also must have the expertise needed to gain and maintain credibility as well as the trust and respect of the workers. In other words, they must be honest and believable. This communication component may be the single most important part of the investigative process. If the facility-management team or the chosen consultants cannot communicate effectively, then someone who can should be brought in to fill this role.

Finally, an environmental evaluation may and often should proceed simultaneously with the medical evaluation. However, it must be recognized that such investigations often identify factors that have no causal relationship to symptoms. For example, an HVAC system may be found to be unbalanced, but that may not be the cause of people's headaches. Only when there is a careful interaction between medical evaluators and environmental investigators can such causal connections be accurately made. And only when results and activities are effectively communicated will problems be resolved with minimal adverse consequences for the organization.

EVALUATING AND CHOOSING CONSULTANTS

Popular concern about the quality of our indoor air has given rise to an explosion of "experts" and consultants, from engineers to duct cleaners to physicians, all ready to proclaim buildings and their occupants sick, to builders ready to charge thousands to millions of dollars to fix things. Entrepreneurship surrounding a new issue of public concern is neither new nor inappropriate. However, when it misrepresents health risks and states of health and produces runaway costs, it demands control. "The building is sick" has become the proclamation of purveyors of expensive, unnecessary services.

Any responsible party—building owners, managers, insurers—confronted with such complaints must exert great caution in selecting a consultant or environmental group to assess the problem. Quite frequently, these consultations and surveys produce more questions than answers and generate data that may appear meaningful but have no toxicological significance. Left uninterpreted, those data may both intensify worker anxieties and contribute to financial liability.

A significant impediment to the effective handling of such complaints arises because of the diversity of professionals involved in the relatively new area of indoor air quality. For example, environmental engineering firms may be prepared to measure substances in indoor air. Lacking toxicological or medical expertise, however, such firms may be ill equipped to interpret the potential public health effects of their findings. Even less frequently are they able to deal with the complaints of specific individuals within that working environment. Because they are neither effective communicators nor health professionals, they cannot respond effectively to the concerns of the workers. This may leave the employer or building manager with a set of data with no meaning, and with no plan of action. It is far easier to collect data than it is to interpret or act upon the information.

Because such investigations involve merging health information and analysis with environmental and engineering assessment, these investigations are of necessity multidisciplinary. Engineers and environmental specialists cannot evaluate health complaints. Medical specialists, without environmental expertise, cannot evaluate the environment. However, because complaints often begin with symptoms, a primary focus of the initial investigation must start with symptom evaluations. Thus, once the situation has demanded the acquisition of outside consultants, those consultants must have sufficient medical expertise to assess symptoms. This is important not only because this method is the most likely to succeed, but also because it minimizes liability. How would it appear to a jury if, after someone claimed that the building caused a serious illness, the building manager had responded by calling an engineer? Ultimately this medical expertise will include physicians, but at first it may be provided by nurses or industrial hygienists. A consulting firm that fails to involve these experts early and that has no readily available medical experts should be avoided.

Ask the environmental firm a few basic questions before hiring it to conduct an IAQ health investigation:

- Who in your organization has medical expertise?
- What physicians do you use?
- How do you define a sick building?
- What do you measure? For each chemical, fungus, mold, or bacterium that you measure, what numbers specifically indicate indoor air problems? What will you compare your numbers to for interpretation?
- How often do you identify indoor air problems and a correctable solution to them?
- What are the normal measured levels of contaminants that you will compare to my building?
- If measured levels of contaminants are in normal ranges, will you then tell me I do not have a sick building?
- Will you meet with the workers to discuss your findings and answer their questions about the health effects? Who will do this?
- Is it possible to satisfy all building occupants?

Those responsible for the quality of building-occupant health and safety must clearly recognize the difference between worker complaints and a proven air-quality problem. Of course you must investigate, but before doing so you must understand the significance of the intended investigation and must question the firm conducting the studies about the significance of potential findings and the expertise of those involved. A key means of assessing consultants' skills is to contact references whose problems the consultants have addressed and solved. Do not embark on a complex exploratory mission of this kind without such a background review.

Case Studies

Case 1

In 1992 air quality consultants forced immediate evacuation of a courthouse building in Florida, proclaiming that mold growth posed a cancer risk to occupants. Built in 1989, the building cost $11 million to construct. The renovation, overseen by those same consultants, cost $9.5 million. Litigation alleged personal injuries of the building occupants and sought to recover the cost of damaged property. It is true that there was mold. There is mold in all buildings in southern Florida. It is true that the building had some structural problems. It is not true that this posed an unusual or immediate threat to the employees as the consultants claimed; nor was it necessary to spend $9.5 million to rebuild this building when far more modest repairs and cleanup would have sufficed. This kind of irresponsible misuse of "expertise" can cost millions of dollars in inappropriate expenditures (as it did in this instance).

The lesson learned: when choosing consultants, it is important to find individuals who think responsibly and use good judgment and common sense to help their clients. Asking "what if" questions will help sort out those who have extreme views from those who are more rational. Getting references is essential. Also, beware of conflicts of interest. Investigators should not profit from remediation.

Case 2

In a school district in central Pennsylvania, an asbestos-abatement program included removal of asbestos floor tiles with a petroleum-based chemical solvent. Following this work, teachers and children noted "chemical" smells, sometimes quite intense, in certain classrooms. The school consulted an engineering firm, which measured levels of specific volatile organic chemicals, pronounced them safe because they were below occupational standards, and departed, leaving teachers and a by now frantic parents' group dissatisfied and more frightened than ever. The engineering group had identified and measured chemicals—but their basis for reassurance was a comparison with industrial settings. The teachers and parents thought this was hardly an apt model for an elementary-school environment.

As concerns mounted, so too did the range of symptoms. Headache, fatigue, and irritation were common. Other complaints included cough, increased frequency of colds, asthma, ear infections, upset stomachs, vomiting, diarrhea, and rashes.

By the time this author got involved, intense emotions had gripped this community, resulting in polarization. The school board was seen as uncaring and accused of covering up a potentially deadly situation. Teachers had mobilized in open revolt, as had parents' groups. Local newspapers and television stations had run stories emphasizing the hazards, the unknowns, the possibilities, and the children's fears. Attorneys were now beginning to enter the scene, offering to represent aggrieved parents in lawsuits on behalf of their children.

As we attempted to sort out the scientific issues from the perceptions, it quickly became clear that doing so was essential but not sufficient. If this situation was to be resolved, the issues as well as the perceptions had to be dealt with. Several facts were apparent. First, some classrooms had a persistent smell of petroleum distillates. Second, there were no carcinogens of concern. Third, measured levels of chemicals were quite low—above the odor threshold, but vastly lower than thresholds of toxic levels either for exposed workers or for young children. Fourth, it was possible that levels of some hydrocarbons were at times sufficiently high to produce some irritant symptoms. It was also quite clear that parents and teachers associated odors with toxicity and that they were absolutely convinced that the school was dangerous.

The best scientific explanation for the symptoms and complaints involved a combination of emotional response to odors, some irritant effects in sensitive individuals, and symptom magnification due to a perceived chemical threat. In other words, the teachers, students, and parents were primed emotionally to associate any and all symptoms and illnesses with that school environment.

Reassuring the parents and teachers that they were not being poisoned was an incomplete solution. Because the odors were so central to the symptoms and the perceptions, they had to be eliminated or at least reduced substantially. Ultimately, the resolution involved a combination of odor reduction and an intense educational effort. It was simply insufficient to compare the levels present with permissible occupational limits. Parents had to be taught basic principles of toxicology: how we know that these levels of chemicals are not going to harm their children, why odor had little to do with toxicity, and how and why symptoms have a variety of explanations. In the end, solving such problems involves addressing both the scientific aspects and the perceptions of toxicity.

This case illustrates two critical points. First, even though the cause of the problem was quite straightforward—more so than usual—critical errors were made. Measurements were performed for substances that were known to be present, since they are components of the petroleum distillate at issue, but that have no known health implications at the very low levels measured. This highlights a key axiom: do not measure things that you are not prepared to talk about. A long list of identified chemicals can be frightening, even when experts realize that the levels are low. Second, the consultants misjudged both the levels of concern and the parental distress occasioned by children at risk. They failed to deal with those concerns directly, and they had no knowledge of children's health issues. This only heightened anxiety and frustration among the parents.

Case 3

Fifteen employees of an accounting firm in Los Angeles sued the developer of the building in which they worked, claiming a variety of building-related ailments. For these injuries they demanded $10 million, but they settled out of court for several million dollars. It was determined that the employees smelled a chemical used to seal the ducts. There were no health risks and no health effects, though the workers claimed otherwise and several physicians supported them.

The problem began as many do, during a major renovation. One compound used in an adjacent office suite was a duct sealant that gave off a variety of strong-smelling volatile chemicals. At the levels involved, no significant or serious health effects were possible, but irritant responses were plausible. It must be remembered that chemical fears run high; thus smells generate intense emotional as well as physical responses. People feel sick because they think they have been poisoned, and they may ascribe all of their ill health (whatever the actual cause) to those same smells. In this case, the chemicals from the sealant probably produced some level of irritation and discomfort. This prompted a call to the local Environmental Protection Agency (EPA) office, which dispatched investigators who arrived fully outfitted in protective equipment and wearing respirators. That frightening scene converted worry into panic and further convinced the employees that they had been poisoned. After all, why else would environmental officials be wearing respirators? The actual reason was that the investigators simply did not know what was present and were taking no chances. They took numerous measurements; no harmful levels of any chemicals were found. But the employees never returned, and they filed lawsuits.

This case illustrates several things. First, and most important, smells lead to problems. It is very important, particularly in these days of heightened awareness and concerns, that renovation be carried out in ways that minimize the potential for exposure. Painting, for example, may be reserved for the weekends and should end early enough for sufficient dilution ventilation to take place before occupants return. When painting must take place while adjacent areas are occupied, measures to minimize exposures to occupants (while providing adequate dilution ventilation to maintain worker protection for the painters) may include the following: exhausting air to the outside using a portable fan while creating negative pressure; sealing all supplies and returns with plastic; modifying the HVAC system to service rooms where the painting will be done; and instituting an effective housekeeping program to control dust (SMACNA, 1995).

Second, psychological factors commonly contribute to associations people make between the workplace and their health. If they believe they have been poisoned, they will feel sick, and it may be impossible to dispel that perception. Also, the belief that one was poisoned is grist for lawsuits. Most of the cases

arising from indoor air claims have at their foundation the claimant's abiding belief that the building caused an illness. And, more often than not, that belief exceeds the reality. In this case, overreaction by environmental-engineering teams,
at the behest of the building manager, contributed to an atmosphere of fear and to
permanent health complaints.

SUMMARY

Confronting indoor air health complaints can be a daunting experience, even for
experienced experts, let alone for a novice building manager or engineer. The
purpose of this chapter was to provide a basic understanding of the causes of
health complaints and the approaches to resolving them. Essential messages to
derive from this discussion include understanding the severity of the situation to
decide when sophisticated help is needed; knowing what to expect of consultants
and how to evaluate them; recognizing the critical importance of both psychological and physical factors; and understanding why common sense, excellent
communication skills, and the ability to react appropriately—neither overreacting nor unreacting—are the keys to a successful resolution.

REFERENCES

Abend, A.C. 1995. A systemic approach to indoor air quality for the building
    manager, In *Occupational Medicine: State-of-the-Art Reviews, Effects of
    the Indoor Environment on Health,* ed. James M. Seltzer, Vol. 10, No. 1
    (January–March), 195–204. Philadelphia: Hanley and Belfus.
Alexander, R.W., and M.J. Fedoruk. 1986. Epidemic psychogenic illness in a
    telephone operators' building. *J Occup Med* 28: 42–45.
American Society for Heating, Refrigeration, and Air Conditioning Engineers
    (ASHRAE). 1981. *Thermal Environmental Conditions for Human Occu-
    pancy.* Atlanta, GA: ASHRAE Standard 55-1981.
American Society for Heating, Refrigeration, and Air Conditioning Engineers
    (ASHRAE). 1989. *Ventilation for Acceptable Indoor Air Quality.* Atlanta:
    ASHRAE Standard 62–1989.
Baker, D.B. 1989. Social and organizational factors in office building-associated
    illness. In *Occupational Medicine: State-of-the-Art Reviews, Problem Build-
    ings: Building-Associated Illness and the Sick Building Syndrome,* ed. James

E. Cone and Michael J. Hodgson, Vol. 4, No. 4, October–December, 607–624. Philadelphia: Hanley and Belfus.

Bauer, R.M., Greve, K.W., Besch, E.L., Schramke, C.J., Crouch, J., Hicks, A., Ware, M.R., and W.B. Lylews. 1992. The role of psychological factors in the report of building-related symptoms in sick building syndrome. *J Con Clin Psych* 60: 213–219.

Brodsky, C.M. 1983. Psychological factors contributing to somatoform diseases attributed to the workplace: The case of intoxication. *J Occup Med* 25: 459–464.

Gots, R.E. 1993. *Toxic Risks: Science, Regulation, and Perception.* Boca Raton, Fla.: Lewis.

Gots, R.E., Gots, B.A., and J. Spencer. 1992. Proving causes of illness in environmental toxicology: "Sick buildings" as an example. *Fresenius Environ Bull* 1: 135–142.

Hall, E.M., and J.V. Johnson. 1989. A case of stress and mass psychogenic illness in industrial workers. *J Occup Med* 31: 243–250.

Hodgson, M.J. 1995. The sick building syndrome: The medical evaluation. In *Occupational Medicine: State-of-the-Art Reviews, Effects of the Indoor Environment on Health,* ed. James M. Seltzer, Vol. 10, No. 1, January–March, 167–176; 177–194. Philadelphia: Hanley and Belfus.

International Society of Indoor Air Quality and Climate (ISIAQ). 1996. *ISIAQ Guideline: Task Force II: General Principles for the Investigation of IAQ Complaints (Draft).* Jan Sundell (Sweden) and Edward Light (USA), Task Force, Co-chairs. ISIAQ: Box 22038, Sub 32, Ottawa, Canada, K1V 0W2 (613-731-2559).

Lees-Haley, P.R. 1993. When sick building complaints arise . . . *Occup Health Safety* 62: 51–54.

Light, E., and J. Tiffany. 1991. Protecting building occupants during constructions and renovations. *Proceedings of IAQ '91.* Washington, D.C.: American Society of Heating, Refrigeration and Air Conditioning.

Persily, A.K. 1994. *Manual for Ventilation Assessment in Mechanically Ventilated Commercial Buildings.* National Institute of Standards and Technology NISTIR 5329.

Quinlan, P., Macher, J.M., Alevantis, L.E., and J.E. Cone. 1989. Protocol for the comprehensive evaluation of building-associated illness. In *Occupational Medicine: State-of-the-Art Reviews, Problem Buildings: Building-Associated Illness and the Sick Building Syndrome,* ed. James E. Cone and Michael J. Hodgson, Vol. 4, No. 4, October–December, 771–797. Philadelphia: Hanley and Belfus.

Samimi, B.S. 1995. The environmental evaluation: Commercial and home. In *Occupational Medicine: State-of-the-Art Reviews, Effects of the Indoor Environment on Health,* ed. James M. Seltzer, Vol. 10, No. 1, January–March, 95–118. Philadelphia: Hanley and Belfus.

Sheet Metal and Air Conditioning Contractors' National Association (SMACNA). November 1995. *IAQ Guidelines for Occupied Buildings under Construction.* Chantilly, Va.: SMACNA.

Spitters, C., Darey, J., Hardin, T., and Ellis, R. January 12, 1996. Outbreak of unexplained illness in a middle school—Washington, April 1994. *MMWR* 45(1): 6–9.

United States Environmental Protection Agency and National Institute of Occupational Safety and Health (US EPA, NIOSH). December 1991. *Building Air Quality: A Guide for Building Owners and Facility Manager.* Washington, D.C.: EPA/400/1-91/033.

United States Environmental Protection Agency. *EPA Indoor Air Quality (IAQ) Clearinghouse.* (800-438-4318).

Ventresca, J.A. January 1995. Ventilation system O&M: A first step for improving IAQ. *ASHRAE Journal* 19–23.

Chapter 4

Indoor Air and Health: Clear-Cut, Equivocal, and Unlikely

Ronald E. Gots, M.D., Ph.D.

Introduction

This chapter addresses the medical and causal links between common symptoms and the phenomenon of indoor air quality (IAQ) concerns. From a medical perspective, several of the health effects people often claim are related to indoor air are not legitimately causal; nor do these effects follow the patterns of clinical diagnosis used in medicine for many years. The author recognizes that well-intentioned and sincere advocates for other views do exist, but this chapter reflects on the findings of medical research papers and the author's own clinical experiences.

The central dilemma in IAQ cases is the differentiation that needs to be drawn between actual diagnosis of clinical illness and the physical complaints that are due to perceived environmental threats. The distinctions between the two are often elusive. IAQ complaints tend to expand beyond the ability of medicine to identify a particular physical cause.

Epidemiologists who track the occurrence of illness have added considerably to medicine's understanding of indoor air illness issues. Although between 500 and 5,000 buildings have been studied for these concerns, relatively few of these investigations have led to clear conclusions that could be implemented into effective corrective action (Stolwijk, 1990). A study by an expert team from Johns Hopkins University's School of Public Health found that identical symptoms were reported from two comparable buildings, one of which had been identified as a "sick" building by some occupants while the other had not. The only difference was that the "sick" building had generated a lot of responses from the occupants. These responses were common complaints such as headaches, but their frequency of attribution to the building was a distinction from the normal rate (Corn, 1991).

Several factors have coalesced in the study of the confusing phenomenon of indoor health effects:

1. More effects can be measured than can be explained. The detectable presence of chemicals and biological contaminants has become more evident as analytical tools have dramatically improved in recent years. Yet the tools have improved faster than our ability to use them in making scientifically defensible risk-management decisions. Some detectable contaminants are relevant to medical evaluations; some are not.

2. Because common symptoms of everyday life such as headaches, fatigue, and nose and eye irritation are the symptoms most likely to be alleged in IAQ concerns, diagnosing the physical problem is simple, but discovering its cause is not. In the patient's mind, the particular building is the culprit. Medical evaluation is made more difficult by such a perception.

3. Solutions to medical related symptoms tend to be diluted according to the number of different disciplines from which the problem solvers are drawn. As these cases have involved toxicology, industrial hygiene, engineering, architecture, public health and manufacturing chemistry issues, the likelihood that solutions will come piecemeal or in divergent directions is much worse than if one "big picture" solution were available. Medical personnel will need to coordinate their work with the efforts made by advisors from other fields.

4. Beliefs outpace data; perceptions become reality. Fear of the effect of a building expands, even if the data to support that belief does not exist or has been actively rebutted by measurements.

Indoor Air's Scientific Debate

The medical community relies on research findings and retrospective evaluations of data. In some buildings, a harmful effect existed and caused illness or death. The classic example is the bacterial contaminant that spread via the ventilation system of the Philadelphia hotel where American Legion members were attending a conference. The spread of the *Legionella* bacteria was a well-studied effect from a clear culprit. The 182 persons who became ill and the 29 who died were in fact affected by a hazardous building.

The debate over building materials and their effects on occupant health gained momentum during the late 1970s, when urea-formaldehyde foam insulation (UFFI) was being challenged. UFFI was eventually banned by the federal

Consumer Product Safety Commission after a congressional inquiry into its risks. The UFFI situation led some members of the public to fear that a tightly sealed building posed excessive chemical exposure risks, although medicine has never conclusively drawn a correlation between the symptoms alleged from residents of homes using UFFI and the causal links of those symptoms.

From the standpoint of medical diagnosis, studies may be made more difficult by the variety of symptoms alleged and causes suggested. In some cases, the first investigator retained as a consultant believes that indoor air is the culprit, and this results in conclusions that the investigator may have been predisposed to discover. Faced with nonspecific complaints and many potential causes, the clinical physician finds that the sheer range of potential causes impedes the ability to do scientific investigation.

Three studies illustrate the difficulties the medical investigator encounters (see a more complete discussion in chapter 3). A Florida courthouse, built in 1989 for $11 million, was evacuated after worker complaints in 1992, and a $9.5 million reconstruction was initiated. The building had structural problems and mold. The decision to evacuate and spend heavily for rebuilding was criticized as an inappropriate response since no immediate or unusual threat existed. Less-drastic repairs and cleanup costs would have sufficed. One lesson learned here is that consultants should not be involved with both evaluation and the more profitable task of remediation, as the consultants were in this case. Another is that checking references and potential conflicts of interest must be part of the evaluation of any consultant.

In a second case, chemical smells were reported inside a school where a petroleum-based chemical solvent was being used to remove asbestos tile. This author became involved after the initial consultant declared the conditions safe because the airborne levels were below occupational-exposure guidelines. Medical complaints increased as parents' concerns mounted; children's symptoms included cough, increased frequency of colds, asthma, ear infections, upset stomachs, vomiting, diarrhea, and rashes.

The lesson was that for this situation to be resolved, perceptions had to be dealt with and treated. Medical advisors must keep in mind that solving odor-related complaints involves *both* the scientific assessment and the perceptions of toxicity. In a school setting, parents felt anxiety and frustration when children's health issues were not directly addressed. Also, measurement should not be done for chemicals for which there is no known health implication; "do not measure things you are not prepared to deal with!" The lay audience may be frightened by a long list of chemicals identified in workplace air, even when experts assure them that levels are low.

A third case is also worth consideration by building managers. A Los Angeles building was undergoing a major renovation. After complaints and litigation, 15 employees of a tenant settled for several millions of dollars in damages.

Odors from a chemical sealant used during renovation volatilized from an adjacent office suite into an accounting firm's offices. Levels of the chemical produced irritation, but no significant or serious health effect was likely. A call to the government for investigation led to an unexpectedly frightening response: government employees who did not know what was present and did not want to risk exposure arrived wearing full respirator-assisted protective gear. The sight of the suited investigators apparently led the employees to fear that serious health risks must exist. Numerous measurements were taken, but no harmful chemical levels were detected. In this case, the government workers' self-protective measures in the face of an unknown airborne vapor were understandable, but the reaction they produced among workers—the increase of fear and alarm—was also understandable.

Some preventive-health lessons from the million-dollar Los Angeles settlements are that renovation of indoor spaces must be undertaken in a way that minimizes the potential for exposure. In areas being painted, for instance, measures such as sealing air supply and return ducts with plastic, temporary heating, venting, and air-conditioning (HVAC) modifications, and exhaust fans to produce negative pressure will reduce the likelihood of odors that generate complaints. Another lesson is that people associate their workplace with safe conditions; if those perceptions change, the responsible officials should recognize and deal with the psychological factors involved. In the Los Angeles building, the arrival of air-sampling investigators in full protective gear with respirators contributed to an atmosphere of fear that generated health complaints.

TERMINOLOGY

The study of indoor air–related disorders is sufficiently new and heterogeneous that the terminology is unclear. "Sick building," "tight building," and "building-related disease" are used interchangeably. Now a new term, *building-related occupant complaint syndrome* (BROCS), has been coined, adding further confusion to this field. Most commonly, the scientific literature places several building-associated conditions into separate categories. These include building-related diseases, tight building or sick building syndrome, and building-associated symptoms. Perhaps BROCS will soon incorporate the latter two expressions.

More recently, new terms describing alleged chronic health effects have arisen. These include multiple chemical sensitivities, toxic encephalopathy, reactive airway dysfunction syndrome (RADS), and occupational asthma.

Building-Related Diseases

Building-related diseases are disorders, ranging from mild to severe, due to specific, identifiable contaminants of indoor air. For a classification of building-related disease to be designated, clear and convincing evidence must exist that something in the building is causal; preferably, the agent should be known. Moreover, the disease or end point of the disorder must generally be quite clear-cut, not merely a set of nonspecific complaints. It may be death or serious respiratory infection, as was the case with Legionnaires' disease. It may be an epidemic of influenza passing through a workforce. It may be an occupational asthma proven by immunological studies of the patient and correlated with cultures of the causative organism, perhaps found in the building's ventilation system.

Legionnaires' disease is an example of a medically determined linkage between death or injury and the defective system in a particular building. There a specific bacterium was causal. Certain other organisms, commonly fungi, molds, and thermophilic bacteria, that contaminate heating and air-conditioning systems produce a variety of complaints and disorders—generally mild hay fever types of allergies, but at times more serious conditions such as asthma or hypersensitivity pneumonia.

Other common infectious diseases, like colds and influenza, may be spread by ventilation systems. In a study comparing Army recruits living in "leaky" barracks to those living in "tight," more energy-efficient barracks, the latter group had a higher frequency of colds (Brundage et al., 1988). When a large percentage of the workforce becomes ill from such infections, building-related diseases can be suspected, although it may be difficult to find a specific contributor in the building environment itself.

It is certainly reasonable to assume that confined spaces with poor outside ventilation would be an environment conducive to the transmittal of respiratory viruses. How and whether that translates to illness in more open and far larger office buildings is, however, not established by medical evidence.

Sick (or Tight) Building Syndrome

The term *sick building syndrome,* or *tight building syndrome,* has been applied to situations in which workers have many and varied symptoms. The sheer range of potential causes of those symptoms makes the term misleading. Hodgson and Cain argued that this term should be abandoned (Cain and Cometto-Muñiz, 1995; Hodgson, 1995a; Hodgson, 1995b). I agree, because the term leads to a false sense that groups of people in offices with symptoms have those symptoms because of some problem with the indoor air quality. However, since the term is so common, I will use it in this chapter.

The term sick building syndrome indicates that people in a workplace either are not feeling well or have health complaints, but it does not explain why. It implies that a significant percentage of building occupants complain of a variety of building-associated symptoms such as eye and mucous membrane irritation, headaches, fatigue, and sinus congestion. Furthermore, it requires a substantial attack rate (involvement by 20 percent or more of building occupants), a temporal relationship to the building, and improvement with specific corrective measures.

This term also implies that problems with indoor air, generally related to poor air exchange in energy-efficient buildings, have been identified. Unlike building-related disease, with sick building syndrome a specific agent such as bacteria or molds is rarely found. It suggests a building-related cause whether or not such a cause exists. People's symptoms may be due to a specific contaminant, but they also might arise from workers' stress or from poor ventilation in an area—what used to be called "stuffy air."

Today, in our body- and health-conscious society, people closely monitor their physical sensations and symptoms; thus symptoms that are merely the result of stuffy air become designated "sick building syndrome." It has been argued, with some merit, that energy-efficient buildings constructed after the early 1970s have sealed internal environments, permitting a variety of contaminants to linger and accumulate when formerly they would have migrated to the outdoors.

While it is clearly true that modern buildings are more tightly sealed, it is not clearly true that indoor air today is worse than it used to be. For example, in this country there are vastly fewer smokers now than in 1965. Conference rooms in office buildings during that era were filled with cigarette and cigar smoke (hence the expression "smoke-filled rooms"). In retrospect, what could be more disturbing to occupants than the hundreds of irritating chemicals emitted by tobacco smoke? That indoor air environment was much more contaminated with secondary tobacco smoke than today's indoor air, whereas today unseen and often unsmelled chemicals are the focus of intense concern. The contrast bears consideration.

MEDICAL INVESTIGATION OF CAUSATION

Because sick building syndrome is associated with nonspecific symptoms and is dependent on subjective individual questionnaires for its identification, its causes—air contamination or psychological factors—cannot easily be distinguished. Other investigators have commented on this (Colligan, 1981). Moreover, as reporting of indoor air problems has become more frequent, there will be increases in psychological influences and reporting biases. The only way to approach some semblance of true scientific investigation is through controlled, blinded studies in which air constituents are varied, unbeknownst to building occupants, and a symptomatology is subsequently reassessed. The few instances in which this was attempted found mixed results regarding the relationship between air-exchange rates and contaminant levels and symptoms (Baldwin and Farant, 1990; Collett et al., 1991; Farant et al., 1990; Farant et al., 1992; Menzies et al., 1990; Menzies et al., 1993; Nagda et al., 1991; Palonen and Seppanen, 1990).

Whether today's sick building syndrome truly represents a defined syndrome is, for all of these reasons, not fully established. The prevalence of complaints alone does not prove that the cause is poor air quality. Complaints could be due to a high pollen count outdoors, a common viral illness, a dissatisfied workforce, reporting bias, or other factors. Only an intense, scientifically controlled investigation might distinguish among these alternatives. Moreover, symptoms do not establish the cause. They typically vary in nature from person to person and are sufficiently nonspecific (having many possible causes) to render uncertain a common causal attribution.

BUILDING-ASSOCIATED SYMPTOMS

The term *building-associated symptoms* is the softest group of building-related conditions or complaints. Here, occupants of a building complain of various symptoms, which they associate with the building. Intensive investigation is unable to elucidate a specific common cause, or the possible cause is too speculative. Much of what has been termed sick building syndrome is probably better called building-associated symptoms.

FACTORS IN THE WORKPLACE THAT CAN PRODUCE SYMPTOMS

Because we are dealing with an eclectic group of symptoms and disorders, their causes are multifaceted, ranging from purely emotional factors to infectious viruses and bacteria. Clearly, as perceptions of bad indoor air increase, emotional factors grow in importance. It becomes increasingly difficult to sort out the real culprits and to separate symptoms due to perceptions from those due to bona fide contaminants. Following is a brief discussion of some of the specific ambient factors in indoor air that may affect levels of comfort and contribute to symptom complaints. We have already discussed factors such as bacteria, which may cause serious diseases, and have considered psychological factors that may cause or intensify complaints.

Ventilation and Related Factors

When researchers seek a cause of sick building syndrome, they most often study the type and quality of building ventilation. But when we read the studies, we see a lot of doubt and little conclusive evidence about whether ventilation alone can explain most cases. The studies vary. The National Institute of Occupational Safety and Health (NIOSH) has reported that ventilation problems existed in at least 53 percent of buildings investigated for indoor air complaints (Wallingford and Carpenter, 1986). That, of course, does not mean these problems caused the complaints.

One frequently cited epidemiological study concludes that air-conditioned buildings consistently show more symptoms than naturally ventilated buildings. Beyond this, no specific cause, such as the use of humidifiers or the presence of formaldehyde or other chemicals, could be identified (Finnegan, Pickering, and Burge, 1984).

Another study concluded specifically that buildings with ventilation from local or central induction fan coil units had more symptoms than buildings with all-air ventilation systems, which in turn had more symptoms than naturally or mechanically ventilated buildings. According to this study, microbiological contamination from chillers, ductwork, or humidifiers (secondary to the ventilation system) can result in some of the worst symptoms, probably by an allergic or endotoxin-related mechanism (Burge et al., 1987).

Another investigator measured a number of environmental characteristics, including thermal parameters (dry-bulb temperature, relative humidity, air speed, and radiant temperature), volatile organic compounds, respirable suspended particulates, lighting and noise intensity, and carbon monoxide and carbon dioxide levels. It was found that certain specific causes stood out; these turned out to be lighting and volatile organic compounds; layers of clothing and crowding were also related to increased symptoms (Hodgson et al., 1991).

The much-quoted Danish study of 4,369 workers in 14 town halls (Skov and Valbjorn, 1987) found no single etiology for sick building syndrome symptoms. Temperature and humidity, carbon dioxide and formaldehyde levels, static electricity, dust, microorganisms, volatile organic compounds (VOCs), and lighting were among the variables studied. Interestingly, the results did not corroborate earlier findings that a higher symptom prevalence exists in mechanically ventilated buildings than in naturally ventilated ones (Robertson et al., 1985).

Other studies done in this country concluded that, even if a single etiology is unknown or there are multiple compounding variables, the type and adequacy of ventilation has more bearing on indoor air quality than any other factor. The specific mechanism by which building ventilation leads to symptoms is unknown; it may be that reduced ventilation directly affects changes in comfort, or it may cause the buildup of chemical pollutants (Letz, 1990). The most likely connection is simply that poorly ventilated air is stuffier and has more odors than well-ventilated air; all this contributes to discomfort and, therefore, to symptoms.

Temperature and Humidity

Temperature and humidity are important comfort factors. Low humidity contributes to dry mucous membranes, which in turn can make the nose and throat feel scratchy, lead to nosebleeds, make throat and nasal membranes more suscep-tible to chemical and other irritants, and contribute to susceptibility to viral in-fections. High humidity can also create health risks, contributing to the growth of biological agents such as fungi.

Temperature affects comfort, and discomfort leads to symptoms. Unfortu-nately, finding the "right" temperature is difficult because some workers will judge a building too hot while others will find it too cold. Air temperature within the building is easy to maintain automatically by means of a thermostat, though an individual's personal comfort also depends on body metabolism and clothing weight. In older buildings, windows and doors can be opened and closed to ad-just airflow. Sealed buildings in which windows cannot be opened have less flex-ibility, so the air-conditioning should be checked to be sure that it is working properly with suitable exchange rates, a balanced delivery system, and no ob-structions in front of vents. Unfortunately, many offices are partitioned in ways that interfere with the original design of the airflow and location of the thermo-stat control; this future flexibility should be kept in mind during the design and construction of new buildings.

Biological Contaminants

Biological contaminants (also called bioaerosols) have received a great deal of attention of late as potential causes of indoor health complaints and disorders. The biological agents at issue are molds (fungi) and bacteria. While it is true that fungi and bacteria can cause certain diseases and symptoms under some circum-stances, it is also true that we live in a world full of biological agents, including fungi and bacteria. The areas most contaminated with mold are dense woods, which are not usually considered threatening.

While some of the concern about mold is justified, in many cases the degree of concern seems to outstrip the actual health threats. Worries about potential, but unproven, health effects have led to extraordinarily expensive remediation of buildings. This attempt to eradicate all mold is both an overreaction and a con-tributor to the popular misperception that all mold is dangerous.

In allergic individuals, high levels of fungi can produce allergic responses such as hay fever or asthma. It is unusual for indoor airborne levels to be high enough to cause such reactions. However, in a heavily mold-contaminated office, it is possible for such responses to develop. Other sources have spent volumes to describe the detailed examination of such responses (Burge, 1995; Cox and Wathes, 1995; Federal-Provincial Committee on Environmental and Occupational Health, 1995). Sometimes these fungi are among the responsible factors in indoor air–related complaints and sick building syndrome, but those occasions are rela-tively unusual—less than 5 percent, according to NIOSH (Melius et al., 1984).

More-serious fungal and bacterial illnesses, such as hypersensitivity pneu-monitis, have also been associated with bioaerosols. For the most part, our experience with such disorders comes from specific occupational exposures that are massive compared with office building exposures. Farmers, silo workers, and mushroom workers are among the susceptible populations.

Relatively few instances of these more-serious lung disorders have been associated with contaminated office buildings. For example, humidifier fever is a true building-related illness caused by biological organisms such as molds and fungi that are traced to standing water in ventilation ductwork and ambient temperature humidification equipment. Its symptoms are wheezing, fluid collection in the lungs, and recurrent fever, which characterize hypersensitivity pneumonitis. Humidifier fever is a respiratory disease that occurs only in sensitive individuals, and it is generally relieved when the affected person leaves the environment. Again, however, such outbreaks are exceedingly rare in the office environment, but they do need to be recognized and managed promptly if they occur.

Recently it has been claimed that certain toxin-producing fungi (particularly *Stachybotrys atra* and various species of *Aspergillus*) when found in an office pose a serious threat (Johanning, Morey, and Jarvis, 1993; Johanning and Yang, 1995; Sorenson, 1990). Although much expensive remediation has resulted from this perception, there is little evidence that the threat is real. True, these agents can cause serious disease) following massive exposures. But there is, to date, no scientific evidence that amounts found on surfaces in offices can give rise to levels that produce harm. Moreover, there is little reason to believe that they can.

It is important to keep in mind that mold on walls is not the same as mold or spores in the air. For the most part, hazardous exposures arise from direct contact with or inhalation of the agents or their spores. Although surface contamination can lead to airborne contamination, the actual quantitative relationship is at best indirect.

Odors

Odors, particularly those that are unfamiliar and are viewed as noxious or "chemical," invoke a wide range of emotional responses (Cometto-Muñiz and Cain, 1993; Knasko, 1993; Cone and Shusterman, 1991; Schiet and Cain, 1990; Shusterman, 1992; Shusterman et al., 1991). Think of responses to the smell of dead flesh. People faint, vomit, and develop palpitations and many other symptoms with such exposures. It is not that the putrefying flesh gives off toxic chemicals, but that it arouses a psychological effect.

Today, chemical smells arouse psychological effects because they are perceived to represent a hazard. Whether or not they actually do is a highly chemical-specific matter. Many completely odorless chemicals (e.g., carbon monoxide) can be quite toxic, whereas other highly odoriferous ones (e.g., mercaptans) are only minimally so. In building-related symptom complexes, however, odors can be extremely and increasingly important.

Odors are probably among the most important causes of health complaints in the indoor environment. Symptom-provoking odors often accompany remodeling or renovation projects; workers smell the chemicals associated with these activities and associate them with risk or danger. The result is symptomatic sick workers.

Emotional Causes and Mass Psychogenic or Sociogenic Illness

The increased awareness of chemicals in the environment and media attention on indoor air issues have bred a greater tendency for psychological factors to aggravate, and even cause, outbreaks of illness in the workplace. Psychosocial issues outside the workplace and stresses within it may result in some of the symptoms, such as headache or lethargy. The use of the term sick building syndrome is an emotionally charged issue already. Research is needed to develop scientific criteria for distinguishing between illness arising from psychological factors and symptoms resulting from exposure to indoor air pollution or toxic substances (Letz, 1990).

Many incidents of epidemic anxiety and mass hysteria have also been reported. These incidents sometimes begin with a remediable indoor air problem; at other times they can be traced not to a building problem but to "problem" individuals. In one acute epidemic, several hundred employees in a state office building in Missouri complained of headache, mucosal irritation, fatigue, odd taste, and dizziness. Extensive investigation revealed no toxic substances or direct cause of the illness. One interesting finding was that the employees who complained of illness were more likely to have perceived unusual odors and inadequate airflow. In any event, investigators concluded that a state of epidemic anxiety was triggered by negative factors in the environment, including poor air quality (i.e., crowding, blocked vents, smoking, high temperatures). Reports of illness from coworkers, arrival of emergency vehicles, and evacuation of the building probably led to the escalation of the event (Donnell et al., 1989).

In another similar incident, operators in a telephone-company building reacted to what they reported was a strange odor with symptoms of headache, nausea, throat irritation, and even respiratory distress. The incident dragged out over an entire month, with evacuations and inspections by California Occupational Safety and Health Administration (OSHA) officials, the local fire department, and the county hazardous materials–management team. No evidence of toxic fumes, gases, chemical leaks, or spills could be found, and all of the people taken to hospitals were found to be asymptomatic, with normal laboratory results. The investigation then turned to one individual who in fact had spread the epidemic as he moved from one part of the building to another with reports of noxious odors that he interpreted as petroleum distillate poisoning.

Here, too, the hysteria escalated as emergency vehicles arrived on the scene and workers witnessed fellow employees in respiratory distress (Alexander and Fedoruk, 1986).

Often, investigations do not reveal a specific causal agent, even with careful monitoring of air contaminants. But because symptoms often disappear with improvements in ventilation or when individuals leave the building, it is difficult to know whether there is a physiological basis for the illness or whether it is a case of mass hysteria or epidemic psychogenic illness.

Products of Combustion, Including Tobacco Smoke

Combustion products are sometimes implicated in building-related illnesses, particularly in cases of respiratory effects. These products consist primarily of carbon monoxide (CO), nitrogen dioxide (NO_2), and sulfur dioxide (SO_2). Sources of these compounds are tobacco smoking, gas ranges, pilot lights, unvented kerosene space heaters, wood and coal stoves, fireplaces, and vehicle emission exhaust.

The contaminant that probably has received the most publicity is environmental tobacco smoke (ETS), also called passive tobacco smoke (PTS), which refers to its effects on nonsmokers. Exposures from passive smoke are mainly sidestream. Though sidestream smoke may have higher concentrations of some toxic and carcinogenic substances than mainstream (active) smoke, it is diluted by room air.

The most consistent and conclusive findings have shown that for children, and particularly for children of parents who smoke, PTS increases the occurrence of respiratory illness and chronic respiratory symptoms such as bronchitis, pneumonia, and coughing (Samet, Marbury, and Spengler, 1987a and 1987b). The health effects of PTS on respiratory symptoms and infection in adults have not been as well studied, and the subject remains controversial. Studies of the association between passive smoking and lung cancer in adults are also inconclusive; case-control and cohort studies do not uniformly indicate increased cancer risk. However, the International Agency for Research on Cancer of the World Health Organization (WHO), the National Research Council, and the United States Surgeon General have all concluded that involuntary smoking is a respiratory carcinogen (Samet, Marbury, and Spengler, 1987a).

Less evidence exists for adverse health effects of some of the other chemical compounds listed here, like sulfur dioxide, partly because they occur infrequently in the indoor occupational environment. For example, though the effects of acute carbon monoxide poisoning by asphyxiation are known, health effects at low levels, and particularly those resulting from chronic exposure, are less well documented. Nitrogen dioxide occurs during combustion of gas during cooking and is emitted from burning pilot lights; exposure is usually residential. The magnitude of respiratory illness resulting from exposure to NO_2 is usually small (Samet, Marbury, and Spengler, 1987a).

Automobile exhaust makes people feel ill because of odors, carbon monoxide, and irritants. At times, building complaints occur where air-intake systems are near or in garages or carports, permitting exhaust fumes to enter the building.

Formaldehyde, VOCs, and Other Chemicals

Medical evaluation of some air complaints is difficult because physicians cannot correlate worker symptoms with the measured levels of chemicals detected in the workplace air. The science of measurement has become very sophisticated, enabling us to detect indoor air contaminants at extremely low levels. Investigators conducting such measurements find hundreds of chemicals around us at all times at these levels. Residential air, building air, outdoor air, air worldwide contains such materials. According to a 1989 report by WHO's Committee on Indoor Air Quality, " . . . the indoor organic air pollutants as reported from several large surveys are similar in the distribution of concentrations in residential environments in several industrialized countries" (WHO, 1989). That study discussed 73 chemicals commonly found in indoor air worldwide.

Thus it is quite easy to identify substances in indoor air. And these substances' complex and frightening organic chemical names (e.g., hexane, formaldehyde, benzene, trichloroethylene, 1,1,1-trichloroethane, methylethylketone) raise concerns in both those who believe that a building is causing their illness and those who are ultimately responsible for the building. Health effects of chemicals at these low levels, however, are often nonexistent or unknown.

For example, a common indoor air complaint is irritation of the nose and eyes. The WHO document previously mentioned noted, "Organic compounds do produce mucosal irritation and other morbidity, though usually at orders of magnitude above the measured concentrations noted indoors." Indoor air concentrations of identified substances are often thousands of times lower than those known to produce health effects. OSHA frequently permits workers in industries that use or manufacture those chemicals exposure to levels one hundred to many thousands of times higher than those found in buildings with no established untoward health effects.

Critics of this disparity point to differences in job requirements, such as intense cognitive functions needed in offices, and differences due to such things as "healthy worker effects" in which industries weed out sensitive workers. Though these and other arguments may be valid, there is little proof that they are.

To some extent, these arguments become rationalizations for people zealously committed to the belief that indoor air in modern buildings is uniformly bad and is producing significant health problems. Thus, frequently levels of chemicals in buildings in which worker complaints have occurred are no higher than customary background levels found in homes, shopping malls, and neighborhood restaurants. Invariably they are vastly lower than permissible exposure levels in manufacturing facilities. This leaves investigators with data that identifies chemicals but cannot correlate those levels of chemicals with the workers' symptoms.

Numerous chemical compounds may contribute to indoor air pollution. These include formaldehyde, asbestos, radon, carbon monoxide, and a category of complex mixtures of VOCs that are typically found in new buildings. Certain questions arise regarding these chemicals and their role in indoor air quality. We need to know in any given instance (a) whether the chemical (or chemicals) is (are) a proven cause of the illness; (b) whether the levels at which the chemicals exist in the environment are known to cause the illness or symptoms; and (c) what scientific methods have been used to measure the chemical levels, document symptoms, and prove causality.

Formaldehyde was used in UFFI until that product was banned; it also has numerous sources in the home and office—particleboard, paper products, floor coverings, and carpet backings are among the sources, albeit in very small amounts. Reported levels of formaldehyde in office buildings have ranged from 0.01 to 0.30 parts per million (ppm), all well below the OSHA standard of 0.75 ppm. Levels that have been measured in buildings with no complaints were typically less than 0.1 ppm, and reports of irritation of the eyes and upper respiratory tract have occurred at levels above 0.1 ppm. The disparity between the OSHA standard and levels of reported complaints should be noted. Formaldehyde has been associated with respiratory and neurobehavioral effects (at lower levels than OSHA permits), but this has not been proven. Published studies have been biased with regard to subject selection and data collection; further investigation is needed (Letz, 1990).

VOCs form a category made up of many different compounds that have been identified in indoor air. As the technology for chemical analysis improves, we have an increasing ability to identify trace amounts of these compounds. In a large-scale series of NIOSH investigations, 350 VOCs were identified in concentrations greater than 0.001 ppm. This does not mean that they exist in greater amounts than they did five years ago, or that exposure is greater, or that there is more danger from them, but simply that our techniques for detecting them are better. In all cases, the measured levels of VOCs were within a factor of 100 of OSHA's permissible exposure levels. Measured levels of VOCs were almost always well below the no-effect levels for acute symptoms in humans (Letz, 1990).

With VOCs, as with other chemicals that may affect IAQ, conflicting or incomplete scientific evidence of toxicity at low levels makes medical evaluation more difficult. One study, for example, found that a small sample of healthy individuals, when exposed for a short time to VOCs, experienced subjective symptoms such as headache and general discomfort but did not show any decreased performance on behavioral tests (Otto et al., 1992).

To the contrary, two other studies found that subjects exposed to organic solvents showed both cognitive deficits and psychological disturbances (similar to posttraumatic stress disorder) on standardized tests. But nowhere in the studies is information given on the levels and intensity of exposure—information we need to compare the levels to those of the same chemicals in indoor air. These studies, then, should not be used as the basis for concluding that low-level VOCs cause neuropsychological disturbance (Molhave, 1992; Molhave, Bach, and Pedersen, 1986; Morrow et al., 1989; Morrow et al., 1990).

CHRONIC ILLNESSES ALLEGED TO BE CAUSED BY ENVIRONMENTAL FACTORS IN OFFICE BUILDINGS

Four potential long-term or chronic illnesses have been blamed on indoor air. Most of these allegations arise in the context of claims of injuries from the workplace. In almost all cases (with a rare exception to be discussed later), these illnesses do not actually occur from office exposures; yet the claims allege otherwise.

One reason that such claims are made is that sick building disorders and building-associated symptoms are self-limited, relatively mild problems that do not create long-term dysfunction or disability. This lack of medical significance does not support the claims of medical damage that are sometimes made in a lawsuit.

Multiple Chemical Sensitivities or Idiopathic Environmental Intolerances

A certain percentage of individuals claim that they are permanently sensitive to all chemicals, and hence disabled, as a result of exposure to something in the workplace. Paints, carpeting, pesticides, copy-machine toner, carbonless copy paper, and standard cleaning chemicals are among the materials that have been blamed by those who view themselves as permanent victims. The claimants are often supported in their belief by a variety of medical practitioners who are equally strong in their beliefs (U.S. EPA, 1996; Gots, 1995). There is, however, no standard or recognized clinical definition for this condition, and no tests, studies, or other objective or reproducible criteria exist with which to make the diagnosis.

In 1985 Dr. Mark Cullen, a professor of occupational medicine at Yale University, named this condition *multiple chemical sensitivities* (MCS) (Cullen,

1987). More recently, a panel of experts convened by a committee of the International Programme on Chemical Safety (IPCS) of WHO and the German Federation of Health and of the Environment met in Berlin, Germany, and renamed this phenomenon *idiopathic environmental intolerances* (IEI) (IPCS, 1996):

Dr. Cullen presented seven criteria for the diagnosis, which included an extremely intense and acute chemical exposure. As practitioners have now promoted or questioned this diagnosis and/or its true cause, many of Cullen's criteria have been overlooked. They are significant and should be applied in any medical evaluation of this type of complaint.

Let us trace the development of a typical MCS/IEI diagnosis. First, the patient tells the physician that chemical exposures make him or her feel sick. Because low levels of chemicals are part of our world and unavoidable, the physician pronounces the individual "sensitive or allergic to the world." These patients have been studied. They do not have allergies. They do not have immune system disorders (despite some claims to the contrary). They have no specific physical abnormalities. Some have serious psychiatric disorders such as depression. Most have the psychiatric condition known as somatization, and many have been influenced by their doctors, who have convinced them, or at least have supported their belief, that they are sensitive to chemicals (Black, 1996).

Because there have been no scientific findings and no logical scientific explanation exists for MCS as a toxic disorder, most major medical and scientific bodies agree that MCS/IEI has not been shown to have a physical basis and is at least as or more likely to have a psychological origin. Such organizations have included the American Medical Association (AMA), the American Academy of Allergy and Immunology (AAAI), the California Medical Association, the American College of Physicians (ACP), the American College of Occupational and Environmental Medicine (ACOEM), the International Society of Regulatory Toxicology and Pharmacology, and a committee of WHO (AAAI, 1986; ACOEM, 1993; ACP, 1989; California Medical Association Scientific Task Force on Clinical Ecology, 1986; AMA, 1992; IPCS, 1996).

MCS might be defined as a phenomenon in which a person develops many symptoms resulting from perceived exposures to low levels of chemicals. There is no question that there are such people. The question, therefore, is not "Does MCS exist?", but "What is MCS?" In almost all instances, a conditioned response can explain the symptoms, just as a person can feel the pain upon walking into the dentist's office or experience distress from the smells of a hospital or an overflowing sewer. These are normal human emotional responses, not toxic effects (Shusterman, 1992; Gots, 1996 [In press]).

Notwithstanding the lack of medical consensus that IEI or MCS is a physical disease, it is a legally significant phenomenon. Many people alleging this disorder have obtained workers' compensation payments or have won judgments against employers and/or building owners.

Toxic Encephalopathy

An individual sometimes claims (usually in the context of a sick building lawsuit) that he or she developed toxic encephalopathy—brain damage—as a result of exposure to chemicals in the workplace. Such an occurrence is scientifically highly unlikely and probably does not occur, but it is worthwhile to understand the basis for this type of claim.

When inhaled at high levels, certain organic solvents may alter brain function acutely or possibly permanently. The best-known examples are, of course, general anesthetics that are used intentionally to alter consciousness and temporarily impair brain function. Chronic alterations of brain function have been categorically shown to occur in individuals who are exposed to exceedingly high levels of certain organic solvents for long periods. For example, glue sniffers who inhale 20,000 to 30,000 ppm of toluene day in and day out for many years eventually develop anatomical alterations in their brains. These abnormalities are identifiable on MRI scans and correlate with neurocognitive dysfunction manifested through neuropsychological testing. This is the classic case of the so-called *psycho-organic syndrome* or *solvent encephalopathy.*

At the lower end of the exposure scale, many studies have examined painters, printers, and others who have been occupationally exposed over long periods to moderately high levels of solvents (although far less than glue sniffers have been). Many of those studies identified subtle neurocognitive alterations identifiable through neuropsychological testing (Hänninen et al., 1976; Hooisma et al., 1993; Olson, 1982; Orbaek et al., 1987). Other studies, however, found conflicting data and did not make such identifications (Bleecker et al., 1991; Cherry et al., 1985; Colvin et al., 1993; Edling et al., 1993). Moreover, all of those studies in which changes were identified involved high exposures over prolonged working lifetimes. No generally accepted and recognized scientific evidence exists for the proposition that low-level or short-term exposures can produce chronic or permanent brain damage.

The reader should be cautious about extrapolations. People who claim that exposures in an office building cause such injuries are extrapolating inappropriately from these high-dose, long-term exposures to chemicals, with a pattern of dosing quite dissimilar to that found in office building exposures. Basic understanding of toxicology tells us that such extrapolations are inappropriate.

Reactive Airways Dysfunction Syndrome and Asthma

RADS and asthma are disorders of the lower airways that make it difficult to move air in and out. Typically, RADS is a disorder caused by exposure to high levels of highly irritating chemicals, such as may occur after a chlorine gas release (Bernstein and Bernstein, 1989; Boulet, 1988; Brooks, Weiss, and Bernstein, 1985a and 1985b). Some have claimed that RADS was caused by irritant substances in indoor air that occurred following such activities as renovation or the installation of new carpeting. There is no convincing evidence that this is true, and such claims arise primarily in the context of litigation. For the most part, levels of irritants from such exposures are far too low to produce RADS.

Asthma differs from RADS in that it is related primarily to environmental allergens as opposed to irritants. Individuals with asthma may be sensitive to agents found indoors, including dust, dust mites (which live in dust), cockroaches, fungi, and mold. Asthmatics are also more sensitive to odors and irritants than nonasthmatics and may experience some discomfort if these are found in the indoor environment. What is not likely true (but has been claimed in the context of building-related lawsuits) is that a person's asthma can be made permanently worse as a result of exposures that produce discomfort or transient exacerbation. Once the building-related exposures cease, the individual is generally no different from before (Chan-Yeung, 1995; Chan-Yeung and Malo, 1995).

Summary

Although health issues that relate to the office environment do exist, it seems that complaints far exceed identifiable environmental causes. This is because building-associated symptoms may or may not have anything to do with indoor air. They may be related to other environmental factors such as lighting, ergonomics, and noise; to psychosocial factors such as job satisfaction, stress, and perceptions of hazards; or to non-work-related factors such as other diseases (i.e., allergies) and home stresses.

Of course, some environmental factors can produce symptoms, such as poor ventilation, odors, infectious agents, molds and fungal spores, and some volatile chemicals. It is important to take complaints seriously and investigate appropriately. It is equally important not to overreact or spend vast sums of money for unnecessary testing or remediation.

The following eight items summarize the information presented in this chapter. Appendix H presents other information for ease of reference.

1. Health issues are most critical when they pose an imminent danger to health or life. Fortunately, such situations in commercial, nonmanufacturing settings are extremely rare. An example is a gas leak that threatens an explosion or asphyxiation. Crises of that magnitude involving bioaerosols have occurred only a couple of times in the history of this country; the outbreak of Legionnaires' disease is the best-known such incident.

2. The psychological aspects of health complaints are as important as the physical ones. In fact, they lead to more litigation and are far more expensive.

3. Building owners can minimize financial risk by being aware of the psychology of indoor health complaints. For example, they can prevent chemical odors from becoming a concern by renovating during periods when people are not in the area.

4. Most health complaints made by people in commercial and public buildings cannot be readily connected to specific environmental problems or findings unless the complaints coincide with painting or other renovations. Invariably, environmental issues can be identified, but only infrequently can they be linked to health effects.

Thus, three questions must be asked:

a. Are there identified environmental problems that need to be corrected?

b. Are people suffering medical problems from building-related factors?

c. Does the level of psychological distress threaten the building and its occupants?

5. For those health complaints that can be related to indoor air problems, the overwhelming majority (99% or more) are minor and pose no serious threat, either short-term or permanent, to individuals. Thus, they can usually be investigated systematically and carefully without undue alarm.

6. Immediate evacuation is *not* usually necessary because of exposure to biological airborne carcinogens (mycotoxins). Such a contention of imminent health risk has no scientific basis and is inappropriately alarming.

7. There is no epidemiological support for or general acceptance of the belief that indoor environmental contaminants in commercial or public nonmanufacturing settings can cause cancer or miscarriages.

REFERENCES

Alexander, R.W., and M.J. Fedoruk. 1986. Epidemic psychogenic illness in a telephone operators' building. *J Occup Med* 28:42–45.

American Academy of Allergy and Immunology (AAA). 1986. Clinical ecology. *J Allergy Clin Immunol* 78:269–271.

American College of Occupational and Environmental Medicine (ACOEM). 1993. ACOEM statement about distinctions among indoor air quality, MCS and ETS. [position statement]

American College of Physicians (ACP). 1989. Clinical ecology. *Ann Intern Med* 111(2):168–178.

Baldwin, M.E., and J.P. Farant. 1990. Study of selected volatile organic compounds in office buildings at different stages of occupancy. In *Proceedings of the Fifth International Conference on Indoor Air Quality and Climate,* Vol. 2, pp. 665–670. Toronto: International Society of Indoor Air Quality and Climate.

Bernstein, I.L., and D.I. Bernstein. 1989. Reactive airways disease syndrome (RADS) after exposure to toxic ammonia fumes. *J Allergy Clin Immunol* 83:173.

Black, D.W. 1996. Case report. Iatrogenic (physician-induced) hypochondriasis: Four patient examples of "chemical sensitivity." *Psychosomatics* 37(4): 390–393.

Bleecker, M., Bolla, K.I., Agnew, J., Schwartz, B.S. and D.P. Ford, 1991. Dose-related subclinical neurobehavioral effects of chronic exposure to low levels of organic solvents. *Am J Ind Med* 19(6):715–728.

Bond, S., Beringer, G.B., Kundin, W.D., et al. 1983. Epidemiologic problems related to medical coverage of new diseases. Presented at the 111th Annual Meeting of the American Public Health Association, November 13–17, 1983, Dallas.

Boulet, L.P. 1988. Increases in airway responsiveness following acute exposure to respiratory irritants: Reactive airway dysfunction syndrome or occupational asthma? *Chest* 94(3):476–481.

Brooks, S.M., Weiss, M.A., and I.L. Bernstein. 1985a. Reactive airways dysfunction syndrome (RADS): Persistent asthma syndrome after high-level irritant exposures. *Chest* 88(3):376–384.

Brooks, S.M., Weiss, M.A., and I.L. Bernstein. 1985b. Reactive airways dysfunction syndrome: Case reports of persistent airways hyperreactivity following high-level irritant exposures. *J Occup Med* 27(7):473–476.

Brundage, J.F., Scott, R., Lednar, W.M., Smith, D.W., and R.N. Miller. 1988. Building-associated risk of febrile acute respiratory diseases in army trainees. *JAMA* 259:2108–2112.

Burge, H.A. 1995. *Bioaerosols, Indoor-Air Research Series.* Boca Raton, Fla.: CRC.

Burge, S., Hedge, A., Wilson, S., Bass, J.H., and A. Robertson. 1987. Sick building syndrome: A study of 4373 office workers. *Ann Occup Hyg* 31(4A): 493–504.

Cain, W.S., and E. Cometto-Muñiz. 1995. Irritation and odor as indicators of indoor pollution. In *Occupational Medicine: State of the Art Reviews: Effects of the Indoor Environment on Health,* ed. James M. Seltzer, Vol. 10, No. 1, January–March, pp. 133–146, Philadelphia: Hanley and Belfus.

Cain, W.S., Samet, J.M., and M.J. Hodgson. July 1995. The quest of negligible health risk from indoor air: ASHRAE's ventilation standard expresses a clear concern for health as well as comfort. *ASHRAE Journal,* pp. 38–43.

California Medical Association Scientific Board Task Force on Clinical Ecology. 1986. Clinical ecology: A critical appraisal. *West J Med* 144:239–245.

Chan-Yeung, M. 1995. ACCP Consensus Statement: Assessment of asthma in the workplace. *Chest* 108(4):1084–1117.

Chan-Yeung, M., and J. Malo. 1995. Review article: Occupational asthma. *N Eng J Med* 333(2):107–112.

Cherry, N., Hutchins, H., Pace, T., and H. Waldron. 1985. Neurobehavioral effects of repeated occupational exposure to toluene and paint solvents. *Br J Ind Med* 42(5):291–300.

Collett et al. 1991. The impact of increased ventilation on indoor air quality. In *Proceedings of IAQ '91: Healthy Buildings.* American Society of Heating, Refrigerating and Air-Conditioning Engineers, pp. 197–200. Atlanta: American Society of Heating, Refrigerating and Air-Conditioning Engineers.

Colligan, M.J. 1981. The psychological effects of indoor air pollution. *Bull NY Acad Med* 57:1014–1026.

Colvin, M., Myers, J., Nell, V., Rees, D., and R. Cronje. 1993. A cross-sectional survey of neurobehavioral effects of chronic solvent exposure on workers in a paint manufacturing plant. *Environ Res* 63:122–132.

Cometto-Muñiz, J.E., and W.S. Cain. 1993. Efficacy of volatile organic compounds in evoking nasal pungency and odor. *Arch Env Health* 48(5): 309–314.

Cone, J.E., and D. Shusterman. 1991. Health effects of indoor odorants. *Environ Health Perspect* 95:53–59.

Corn, M. 1991. Personal communication.

Council on Scientific Affairs, American Medical Association (AMA). 1992. Council report: Clinical ecology. *JAMA* 268(24):3465–3467.

Cox, C.S., and C.M. Wathes, eds. 1995. *Bioaerosols Handbook.* Boca Raton, Fla.: CRC Press, 1995).

Cullen, M.R. 1987. The worker with multiple chemical sensitivities: An overview. In *Occupational Medicine: State of the Art Reviews: Workers with Multiple Chemical Sensitivities,* ed. Mark R. Cullen, Vol. 2, No. 4, October–December, pp. 655–661. Philadelphia: Hanley and Belfus.

Daniell, W., Camp, J., and Horstman, S. 1991. Trial of a negative ion generator device in remediating problems related to indoor air quality. *J Occup Med* 33:681–687.

Donnell, H.D., Bagby, J.R., Harmon, R.G., Crellin, J.R., Chaski, H.C., Bright, M.F., Van Tuinen, M., and R.W. Metzger. 1989. Report of an illness outbreak at the Harry S. Truman State Office Building. *Am J Epidemiol* 129: 550–558.

Edling, C., Anundi, H., Johanson, G., and K. Nilsson. 1993. Increase in neuropsychiatric symptoms after occupational exposure to low levels of styrene. *Br J Ind Med* 50:843–850.

Farant, J.P., et al. 1990. Effect of changes in the operation of a building's ventilation system on environmental conditions at individual workstations in an office complex. In *Proceedings of the Fifth International Conference on Indoor Air Quality and Climate,* Vol. 1, pp. 581–585. Toronto: International Society of Indoor Air Quality and Climate.

Farant, J.P., et al. 1992. Environmental conditions in a recently constructed office building before and after the implementation of energy conservation measures. *Appl Occup Environ Hyg* 7:93–100.

Federal-Provincial Committee on Environmental and Occupational Health. 1995. *Fungal contamination in public buildings: A guide to recognition and management,* pp. 9–28, 59–76. Ottawa: Health Canada.

Finnegan, M.J., Pickering, C.A.C., and P.S. Burge. 1984. The sick building syndrome: Prevalence studies. *Br Med J* 289:1573–1575.

Fraser, D.W., Tsai, T.R., Ornstein, W., Parkin, W.E., Beecham, H.J., Sharrar, R.G., Harris, J., Mallison, G.F., Martin, S.M., McDale, J.E., Shepard, C.C., Brachman, P.S., and the field investigative team. 1977. Legionnaires' disease: Description of an epidemic of pneumonia. *N Engl J Med* 297: 1189–1197.

Gots, R.E. 1995. Multiple chemical sensitivities—public policy. *Clin Tox* 33(2): 111–113.

Gots, R.E. 1996. MCS: Distinguishing between psychogenic and toxicodynamic. *Reg Tox Pharm* 24: S8–S15.

Hänninen, H., Eskelinen, L., Husman, K., and M. Nurminen. 1976. Behavioral effects of long-term exposure to a mixture of organic solvents. *Scand J Work Environ Health* 2(4):240–255.

Hedge, A. 1988. Job stress, job satisfaction and work-related illness in offices. In *Proceedings of the 32nd Annual Meeting of the Human Factors Society,* Vol. 2, pp. 777–779, Anaheim, California.

Hodgson, M.J. 1989. Clinical diagnosis and management of building-related illness and the sick-building syndrome. In *Occupational Medicine: State of the Art Reviews, Problem Buildings: Building-Associated Illness and the Sick Building Syndrome,* ed. James E. Cone and Michael S. Hodgson, Vol. 4, No. 4, October–December, pp. 771–797. Philadelphia: Hanley and Belfus.

Hodgson, M.J. 1995a. The medical evaluation. In *Occupational Medicine: State of the Art Reviews: Effects of the Indoor Environment on Health,* ed. James M. Seltzer, Vol. 10, No. 1, January–March, pp. 177–194. Philadelphia: Hanley and Belfus.

Hodgson, M.J. 1995b. The medical evaluation. In *Occupational Medicine: State of the Art Reviews, Effects of the Indoor Environment on Health,* ed. James M. Seltzer, Vol. 10, No. 1, January–March, pp. 177–194. Philadelphia: Hanley and Belfus.

Hodgson, M.J., Frohliger, J., Permar, E., Tidwell, C., Traven, N.D., Olenchock, S.A., and M. Karpf. 1991. Symptoms and microenvironmental measures in nonproblem buildings. *J Occup Med* 33(4):527–533.

Hooisma, J., Hänninen, H., Emmen, H.H. and B.M. Kulig. 1993. Behavioral effects of exposure to organic solvents in Dutch painters. *Neurotoxicol Teratol* 15:397–406.

Hopwood, D.G., and Guidotti, T.L. 1988. Recall bias in exposed subjects following a toxic exposure incident. *Arch Environ Health* 43:234–237.

International Programme on Chemical Safety (IPCS). 1996. Conclusions and recommendations of a workshop on 'multiple chemical sensitivities (MCS)' 21–3. February 1996, Berlin. Organized in collaboration with the International Programme on Chemical Safety (UNEP-ILO-WHO), the Federal Ministry of Health (BMG), the Federal Institute for Health Protection of Consumers and Veterinary Medicine (BLVV), and the Federal Environmental Agency (UBA).

Johanning, E., Morey, P., and B. Jarvis. 1993. Clinical-epidemiological investigation of health effects caused by *Stachybotrys atra* building contamination. In *Proceedings of Indoor Air '93*, July 4–8, Vol. 1, pp. 225–230, Helsinki, Finland.

Johanning, E., and C.S. Yang, ed. 1995. *Fungi and bacteria in indoor air environments: Health effects, detection and remediation.* Proceedings of the International Conference, Saratoga Springs, New York, October 6–7, 1994. Latham, N.Y.: Eastern New York Occupational Health Program.

Knasko, S.C. 1993. Performance, mood and health during exposure to intermittent odors. *Arch Environ Health* 48(5):305–308.

Letz, G.A. 1990. Sick building syndrome: Acute illness among office workers: The role of building ventilation, airborne contaminants and work stress. *Allergy Proc* 11:109–116.

McDade, J.E., Shepard, C.C., Fraser, D.W., Tsai, T.R., Redus, M.A., Dowdle, W.R., and the laboratory investigation team. 1977. Legionnaires' disease: Isolation of a bacterium and demonstration of its role in other respiratory diseases. *N Engl J Med* 297:1197–1203.

Melius, J., Wallingford, K., Kennlyside R., and J. Carpenter. 1984. Indoor air quality: the NIOSH experience. *Ann Amer Cong Govt Ind Hygienists* 10:3–7.

Menzies, R., Tamblyn, R., Farant, J., Hanley, J., Nunes, F., and R. Tamblyn. 1993. The effect of varying levels of outdoor-air supply on the symptoms of sick building syndrome. *N Engl J Med* 328(12):821–827.

Menzies, R.I., et al. 1990. Sick building syndrome: The effect of changes in ventilation rates on symptom prevalence: The evaluation of a double blind experimental approach. In *Proceedings of the Fifth International Conference on Indoor Air Quality and Climate,* Vol. 1, pp. 519–524. Toronto: International Society of Indoor Air Quality and Climate.

Molhave, L. 1992. Controlled experiments for studies of the sick building syndrome. *Annals of the New York Academy of Sciences: Sources of Indoor Contaminants,* ed. W.G. Tucker, B.P. Leaderer, L. Molhave, and W.S. Cain, Vol. 641, pp. 46–55. New York: New York Academy of Sciences.

Molhave, L., Bach, B., and O.F. Pedersen. 1986. Human reactions to low concentrations of volatile organic compounds. *Environment International* 12:167–175.

Morrow, L.A., Ryan, C.M., Hodgson, M.J., and M. Robin. 1990. Alterations in cognitive functioning after organic solvent exposure. *J Occup Med* 32: 444–450.

Morrow, L.A., Ryan, C.M., Goldstein, G., and M.J. Hodgson. 1989. A distinct pattern of personality disturbances following exposure to mixtures of organic solvents. *J Occup Med* 31:743–746.

Nagda, N.L., et al. 1991. Effect of ventilation rate in a healthy building. In *Proceedings of IAQ '91: Healthy Buildings.* American Society of Heating, Refrigerating and Air-Conditioning Engineers, pp. 101–107. Atlanta.

New Jersey Department of Health, Environment Health Hazard Evaluation Program (NJDOH-EHHEP) prepared in cooperation with Atlantic County Health Department). 1983. *Health Survey of the Population Living Near the Price Landfill, Egg Harbor Township, Atlantic County.* New Jersey Department of Health.

Olson, B.A. 1982. Effects of organic solvents on behavioral performance of workers in the paint industry. *Neurobehav Toxicol Teratol* 4(6):703–708.

Orbaek, P., Lindgren, M., Olivecrona, H., and B. Haeger-Aronsen. 1987. Computed tomography and psychometric test performances in patients with solvent induced chronic toxic encephalopathy and healthy controls. *Br J Ind Med* 44(3):175–179.

Otto, D.A., Hudnell, N.K., House, D.E., Molhave, L., and W. Counts. 1992. Exposure of humans to a volatile organic mixture. I. Behavior assessment. *Arch Env Health* 47:23–30.

Palonen, J., and O. Seppanen. 1990. Design criteria for central ventilation and air-conditioning system of offices in cold climate. In *Proceedings of the Fifth International Conference on Indoor Air Quality and Climate,* Vol. 4, pp. 299–304. Toronto: International Society of Indoor Air Quality and Climate.

Robertson, A.S., Burge, S., Hedge, A., Sims, J., Gill, F.S., Finnegan M., Pickering, C.A.C., and G. Dalton. 1985. Comparison of health problems related to work and environmental measurements in two office buildings with different ventilation systems. *Br Med J* 291:373–376.

Robertson, A., and S. Burge. 1986. Building sickness . . . all in the mind. *Occup Health Safety* 38(3):78–81.

Samet, J.M., Marbury, M.C., and J.D. Spengler. 1987a. Health effects and sources of indoor air pollution. Part I. *Am Rev Resp Dis* 136:1486–1508.

Samet, J.M., Marbury, M.C., and J.D. Spengler. 1987b. Respiratory effects of indoor air pollution. *J Allergy Clin Immunol* 79:685–700.

Schiet, F.T., and W.S. Cain. 1990. Odor intensity of mixed and unmixed stimuli under environmentally realistic conditions. *Perception* 19:123–132.

Shusterman, D. 1992. Critical review: The health significance of environmental odor pollution. *Arch Environ Health* 47:76–87.

Shusterman, D., Lipscomb, J., Neutra, R., and K. Satin. 1991. Symptom prevalence and odor-worry interaction near hazardous waste sites. *Environ Health Perspect* 94:25–30.

Skov, P., Valbjorn, O., and the Danish Indoor Climate Study Group. 1987. The 'sick' building syndrome in the office environment: The Danish Town Hall Study. *Environ Int* 13:339–349.

Sorenson, W.G. 1990. Mycotoxins as potential occupational hazards. *J Ind Microbiol, Suppl. 5* 31:205–211.

Stolwijk, J.A.J. 1990. Unpublished report. The "sick building" syndrome. Department of Epidemiology and Public Health, Yale University School of Medicine, New Haven, Conn.

United States Environmental Protection Agency (US EPA). June 18, 1996. Hazardous air pollutant list; modification. Caprolactam: Final rule. *Federal Register* 61(118):30816–30823.

Wallinford, K.M., and J. Carpenter. 1986. In *Proceedings of IAQ '86: Managing Indoor Air for Health and Energy Conservation.* Atlanta: American Society of Heating, Refrigeration and Air Conditioning Engineers.

Woods, J.E. 1989. Cost avoidance and productivity in owning and operating buildings. *Occup Med* 4:753–770.

Woods, J.E., Drewry, G.M., and P.R. Morey. 1987. Office worker perceptions of indoor air quality effects on discomfort and performance. In *Indoor Air '87, Proceedings of the 4th International Conference on Indoor Air and Climate,* ed. B. Seifert, H. Esdorn, M. Fischer, H. Ruden, and J. Wegner. Berlin: Institute for Water, Soil and Air Hygiene.

World Health Organization (WHO) Regional Office for Europe, Copenhagen. 1989. *Indoor Air Quality: Organic Pollutants, Report on a WHO Meeting, Berlin (West),* August 1987, pp. 23–27.

Chapter 5

What Can We Learn about Indoor Environmental Quality Concerns from Studies?

ALAN HEDGE, PH.D.

OVERVIEW

This chapter reviews scientific literature on indoor environmental quality (IEQ) issues, sometimes called sick building syndrome (SBS), in office buildings. Epidemiological research on SBS began in Europe in the early 1980s. Since then, studies of indoor environmental health in office buildings have been conducted in many countries. Findings from these studies are summarized in the following sections.

FACTORS CAUSING IEQ CONCERNS

Types of Ventilation

European and United States research shows that somehow, air-conditioning of an office building is a risk factor for IEQ concerns. This was first suggested in the United Kingdom by Finnegan et al. (1984), who conducted an interview study that compared self-reports of SBS symptoms among workers in nine office buildings and found evidence that symptoms were more prevalent among those workers occupying air-conditioned buildings than those in naturally ventilated buildings. In air-conditioned buildings the air may pass through several potential sources of contamination (e.g., ducts, humidifier, chiller), whereas in naturally ventilated buildings ventilation depends solely on operable windows and doors.

In a survey of workers from six office buildings, Hedge (1984) found differences in the prevalence of eye, nose, and throat symptoms between air-conditioned and naturally ventilated offices. Reports of headache did not differ between offices as a function of type of ventilation, but they did vary as a function of office layout, with the greatest prevalence being reported in deep, open-plan office spaces.

Robertson et al. (1985) compared symptom reports for workers in adjoining air-conditioned and naturally ventilated office buildings. SBS symptoms were more prevalent among workers in the air-conditioned offices than among their counterparts in the naturally ventilated offices. Measurement of a variety of physical environmental parameters (air temperature, globe temperature, air velocity, negative and positive air ion concentrations, formaldehyde, carbon monoxide, and ozone) were significantly different in the two buildings. Hedge et al. (1989a) also compared symptom reports and environmental conditions for workers in the same organization occupying adjacent air-conditioned and naturally ventilated office buildings. Significant differences in SBS symptom reports between the two buildings were found but few differences in indoor air quality, apart from higher concentrations of total volatile organic compounds (TVOC) in the air-conditioned building. A significant association between formaldehyde levels and reports of employees leaving work early because they felt unwell was found for both buildings.

Burge et al. (1987), Wilson and Hedge (1987), and Hedge et al. (1989b) conducted a nationwide questionnaire survey of 4,473 office workers in 47 office buildings in the United Kingdom. Their results confirm an overall difference in SBS symptom prevalence between air-conditioned and unconditioned offices. Symptoms were, however, less prevalent in mechanically ventilated buildings (i.e., those with ducted-air or forced-air extract but with no conditioning of the supply air) than in naturally ventilated buildings. Symptoms were more prevalent in buildings with air-conditioning, but among the air-conditioned buildings, symptoms were generally less prevalent in variable air volume heating, ventilating, and air conditioning (HVAC) systems (i.e., those with ceiling induction or fan-coil units in which air is ducted from a central plant room and then either heated or cooled in the room), or local HVAC systems (i.e., those with no ducting but where air is directly supplied to the office from the outside by wall-mounted units without any ducts). They also found that a variety of nonenvironmental factors were associated with symptom reports (see "Indoor Air Pollutants" later in this chapter). Unfortunately, this survey did not measure physical environment conditions, and to what extent the symptoms were a consequence of exposure to indoor air pollutants is unclear.

A follow-up study (Wilson et al., 1987) surveyed six of the offices from the earlier study some two years after the original work. Results showed similar levels of SBS symptoms in the buildings (the correlation coefficient between BSS[1] for the first and second studies was 0.96). In this follow-up work, environmental conditions and worker surveys were measured in each building. Results showed a significant positive correlation between relative humidity and SBS symptoms, but this was the reverse of what was expected, that is, higher symptom reports at higher humidities and lower symptom reports at lower humidities. There was no relationship between SBS symptoms and air temperature, globe temperature, carbon dioxide concentrations, design air change rates per hour, or occupant density/m[2]. There was a tendency for a negative association between adverse environmental conditions and SBS symptoms, that is, the better the conditions, the worse the symptoms. Nonenvironmental influences on SBS symptoms were also found (see "IEQ Concerns and Nonenvironmental Variables" later in this chapter).

Harrison et al. (1987) surveyed 2,587 workers in 27 office buildings and showed that SBS symptoms were significantly more prevalent in air-conditioned buildings compared to naturally ventilated buildings. Subsequently, Harrison et al. (1990) studied 13 office buildings and found a negative correlation between the number of SBS symptoms and total airborne particulates when the data were analyzed across ventilation classes, but a positive correlation between total airborne particulates and SBS symptoms when the data were analyzed within a ventilation class.

Skov et al. (1988) surveyed 4,369 office workers in 14 Danish town halls and 14 nearby office buildings; at the same time they conducted extensive analyses of the physical conditions and indoor air quality (IAQ) in these buildings. Their results showed considerable variation in the prevalence of SBS between buildings but no difference in symptom prevalence between mechanically ventilated and naturally ventilated buildings (Valbjørn and Skov, 1987; Skov et al., 1990). A recent reanalysis of these data (Mendell and Smith, 1990) has shown that only one of the mechanically ventilated Danish buildings was, in fact, air-conditioned, and when the data are reanalyzed to take account of this, they fit the pattern of the United Kingdom data reported by Finnegan et al. (1984) and Burge et al. (1987). The Danish results also showed that none of the indoor air pollutants measured correlated with symptom reports, although symptoms were correlated with some of the physical environment measures: a fleece factor (the area of material surface divided by the room volume), a shelf factor (the area of open shelving divided by the volume of the room), total dust, and air temperature. They also found that the highest prevalence of SBS symptoms was in the building with the highest level of TVOC. A number of nonenvironmental influences on SBS symptoms were found (see "Factors Causing IEQ Concerns").

---

[1] The building sickness score (BSS) is the average number of SBS symptoms per worker in a building (Burge et al., 1992, 1993).

Mendell et al. (1996) and Fisk et al. (1993) reported a study of 12 northern California office buildings. They found that SBS symptoms were more prevalent in air-conditioned and mechanically ventilated buildings than in naturally ventilated buildings. Jaakkola et al. (1993) also found lower symptom prevalence rates in naturally ventilated than in mechanically ventilated buildings.

Research results have consistently shown an increase in the risk of SBS in air-conditioned buildings. What is it about air-conditioning in a building that might underlie this increase?

Ventilation Rate

Two aspects of ventilation might affect SBS symptoms: ventilation rate[2] and percentage of outdoor air.

Several investigators have tested whether SBS symptoms arise because of inadequate outdoor air ventilation. In a Canadian study, Sterling and Sterling (1983) compared symptoms for workers in a building that was ventilated with either 25 percent or 100 percent outdoor air. Almost no differences in symptom prevalence were found between the two levels. A Finnish study found no difference in the prevalence of SBS symptoms in a building with either 100 percent or 25 percent outdoor air (Jaakkola et al., 1990). Also, Jaakkola et al. (1990) found no differences in SBS symptoms between two identical buildings, one with 0 percent recirculated air and the other with 70 percent recirculated air.

Other researchers have focused on studying ventilation rate and SBS symptoms. A U.S. study of a 20-story government office building compared seasonal effects (winter and summer) of ventilation at either 20 cfmpp or 35 cfmpp. Results showed that, especially in summer, the prevalence of some health symptoms was significantly higher at the lower ventilation rate. Other studies have not confirmed this result. In a double-blind study, Menzies et al. (1990, 1993) found that increasing ventilation provision to workers from 10 through 20 to 50 cfmpp did not reduce SBS complaints. In a Swedish study, Wyon (1992) compared lower (about 23 cfmpp) and higher (about 33 cfmpp) ventilation rates and found no differences in symptom prevalence. However, ventilation rates substantially below 20 cfmpp may be associated with an increased risk of SBS (Mendell, 1993). For the interested reader, Godish and Spengler (1996) provide an excellent, comprehensive review of these studies.

---

[2] Ventilation rate refers to the amount of air supplied per person per unit of time; this is expressed as the number of cubic feet per minute per person (cfmpp). The United States uses imperial measures (cfmpp). Other countries use the metric measure of liters per second per person (L/s. person). A ventilation rate of 20 cfmpp is equivalent to 10 L/s. person.

Indoor Air Pollutants

Evidence on the associations between IEQ, or SBS, and physical environmental conditions generally is sparse. Jaakkola et al. (1989) report a linear relationship between the total number of SBS symptoms and an increase in air temperature above 22°C, but this has not been consistently confirmed in other studies. Helsing et al. (1987) found that increasing the ventilation rate seemed to reduce complaints of SBS symptoms, even though concentrations of carbon dioxide and other indoor air pollutants were all below the current American Society of Heating, Refrigerating, and Air Conditioning Engineers (ASHRAE) standard (ASHRAE, 1989).

Contrary to Danish findings reported by Skov et al. (1989), Sundell et al. (1993) reported a negative correlation between TVOC and SBS symptoms. Norbäck et al. (1989) studied the prevalence of SBS symptoms among different occupational groups: hospital and industrial workers who were exposed to airborne gaseous irritants and office workers who, according to hygiene measures, were not exposed to comparable levels of these irritants. They found that SBS symptoms were more prevalent among the office workers, even though chemical exposures were lower for this group. They also found that women reported more symptoms than men, but that individual factors (i.e., mean age, atopy frequency, and smoking habits) and occupational factors (i.e., work stress, work satisfaction) were not related to symptom reporting. However, in this study SBS symptoms were not defined using a work-related criterion, and the confounding effects of this are unclear.

Hodgson et al. (1987) found that SBS symptom reports were correlated with vibration from nearby mechanical ventilation equipment. This has not been found in other SBS investigations. To help improve IAQ, many U.S. buildings have prohibited cigarette smoking or restricted this to specific areas. In a study of 27 office buildings with either variable air volume or constant air volume ventilation systems, environmental tobacco smoke was not found to substantially affect IAQ in office areas or correlate with the prevalence of SBS (Hedge et al., 1994).

Armstrong et al. (1989) investigated SBS symptoms in a high-rise public office building and found that symptoms were associated with total suspended particulates but not with any other indoor air pollutants. Similarly, in a detailed study of microenvironmental conditions in a "sick" building, Hodgson and Collopy (1989) and Hodgson et al. (1991) found that symptoms did not correlate with any IAQ measures apart from respirable suspended particulates. Wallace et al. (1991) found that SBS symptoms correlate with the perceived dustiness of the office, but unfortunately, levels of office dust were not measured.

Recently, other significant associations between dust or biological contaminants and SBS have been reported. Hedge et al. (1993b) found a high correlation between mineral fibers in settled office dust and SBS symptoms; Gyntelberg et al. (1996) found correlations between SBS symptoms and allergenic materials, fibers, and TVOC in settled dust. The Danish Town Hall study found correlations between SBS symptoms and macromolecular organic dust (MOD), which is the allergenic fraction of settled dust (Skov et al., 1989; Gravesen et al., 1993). Teeuw et al. (1994) found a significant association between gram-negative rod bacteria and airborne endotoxin. Exposure to molds can also evoke SBS-like symptoms. Volatile organic compounds produced by microbiological growth (MVOCs) can be used to detect unseen mold growth in buildings (Strom et al., 1993). Norback et al. (1995) reported that MVOCs can be detected in water-based latex paints used in buildings.

The possibility that dust and its associated contaminants might influence SBS reports also receives some support from a study by Hedge et al. (1993b) that found that installing a high-efficiency air-filtration system into the office improved perceived environmental conditions and health reports. Also, Leinster et al. (1990) and Raw et al. (1993) found that thorough office cleaning is an effective way of reducing SBS symptom rates.

The published literature does not support an association between routine use of pesticides and symptoms of indoor environmental health concern. The nature of SBS allegations, which are alleviated when affected individuals leave a building and return upon reoccupation, do not show the symptoms of pesticide exposure. Pesticide exposures would show some progressive longer-term dysfunction that is not linked to a particular setting. Chapter 4 discussed medical investigation, but the literature does not support claims that SBS has been caused by routine use of pesticides. In the event that a health effect is alleged to have arisen from pesticide exposures, the U.S. Environmental Protection Agency's (EPA) pesticide databases can provide additional information. Appendix C provides information for contacting the EPA.

IEQ CONCERNS AND NONENVIRONMENTAL VARIABLES

There is considerable agreement among studies regarding evidence for several nonenvironmental influences on symptom prevalence. These include personal factors such as gender and occupational factors such as use of a computer and job stress (Burge et al., 1987; Hedge, 1988; Hedge et al., 1989a; Skov, Valbjørn, Pederson, and Danish Indoor Study Group, 1989; Tamblyn and Menzies, 1992). Zweers et al. (1992) surveyed 7,000 Dutch workers in 61 offices and also found that workers' gender, job satisfaction, and video display terminal (VDT) use were among the most significant correlates of symptom reports. Although workers associated their symptoms with environmental conditions, daily fluctuations

in temperature and humidity and reported symptoms of dry nose or nasal congestion were not correlated, but SBS symptoms correlated with self-reported work
stress (Morris and Hawkins, 1987).

In a survey of workers from six office buildings, Hedge (1984) found that
self-reported health symptoms (headache; lethargy; and eye, nose, and throat irritation) were significantly more widespread among women than men in these offices. Gender differences[3] in SBS symptom reporting (more symptoms having
been reported by more women than men) subsequently have been confirmed in
many studies (Burge et al., 1987; Hedge et al., 1989a, 1989b, 1992, 1993a, 1995,
1996; Lenvik, 1990; Skov et al., 1989; Stenberg and Wall, 1995; Tamblyn et al.,
1993; Wilson and Hedge, 1987; Zweers et al., 1990, 1992).

Burge et al. (1987), Wilson and Hedge (1987), and Hedge et al. (1989a)
have shown that in addition to SBS reporting being influenced by individual factors (gender, age, perceived control, attitudes to environmental conditions),
symptom reports are also influenced by occupational factors (computer use, job
stress), and organizational factors (organization type, office type). Other studies
have also found that SBS symptoms are significantly associated with occupational factors such as job grade, hours of VDT use, handling of carbonless copy
paper, and photocopying (Skov et al., 1989). Zweers et al. (1990, 1992) surveyed
more than 7,000 Dutch workers in 61 offices and also found that in addition to
gender, job satisfaction and computer use were among the most significant correlates of symptom reports.

In a study of 764 workers in 3 office buildings, Lenvik (1987) found that the
prevalence of SBS symptoms did differ by workers' gender but did not differ by job
type or job satisfaction. However, men reported greater job satisfaction, and this
may have confounded gender results. In a survey of more than 2,000 U.S. office
workers by Klitzman and Stellman (1989), correlations between perceived IAQ and
psychological well-being as well as between psychosocial job factors and health
symptoms were found, although symptoms were not exclusively those of SBS.

Job stress and workers' negative perceptions of environmental conditions
significantly correlate with SBS symptoms (Hedge, 1988). Several studies have
also found significant correlation between SBS symptoms and either job stress or
job satisfaction[4] (Burge et al., 1987; Hedge, 1988; Hedge et al., 1989a; Hodgson

---

[3] Norback (1990) found that adjusting results for differences in nickel allergy, hyperreactivity,
and infection proneness rendered gender differences in symptom reports statistically insignificant. However, the symptoms studied were not defined as work-related (to be work-related a
symptom must occur when an individual is at work and be alleviated when away from work);
this result has yet to be confirmed in other studies.

[4] Norback et al. (1989) found that SBS symptoms were not associated with job stress or job satisfaction. Unfortunately, in this study, job stress and job satisfaction were measured on a single-
item, analog rating scale, whereas most other recent studies have used multiple-item fixed-
point rating scales, and SBS symptoms were not defined as work-related.

et al., 1992; Skov and Valbjørn, 1989; Tamblyn and Menzies, 1992). Although workers associated their symptoms with environmental conditions, daily fluctuations in temperature and humidity and reported symptoms of dry nose or nasal congestion were not correlated, but SBS symptoms did correlate with self-reported work stress (Morris and Hawkins, 1987).

Eriksson et al. (1996), reporting on the Office Illness Project in northern Sweden, gave a detailed description of occupational influences on SBS. They confirmed that SBS appears to be multifactorial in origin. Symptom reports were influenced by worry about changes in work and the workplace; work demands, especially a large amount of work (workload); perceived role conflict at work; and overall workplace satisfaction. They did not, however, find any association between SBS and underutilization of work skills or control over work.

A U.S. study of 3,155 workers in 18 private-sector offices with generally comparable IAQ found that job stress, job satisfaction, computer use, and gender all exert stronger influences on reports of SBS symptoms than do gaseous indoor air pollutants (Hedge et al., 1992). An extension of this survey that included another 9 buildings and more than 1,000 additional workers for a total of 27 buildings and more than 4,000 workers, reinforced the findings that personal and occupational factors are significantly associated with SBS (Hedge et al., 1994, 1995, 1996).

FACTORS CAUSING SBS: A MULTIFACTORIAL MODEL

The evidence reviewed suggests that SBS symptoms are not caused simply by direct exposure to indoor air pollutants in offices, but rather they arise as the consequence of exposure to multiple risk factors. Some of these may be airborne, some are environmental conditions, and some are personal and occupational.

Hedge (1990) proposed that any adverse effects of the office environment on occupant comfort and health may arise from *direct environmental effects,* for example, exposure to pollutants; *indirect environmental effects,* for example, a worker's satisfaction with thermal conditions; and *nonenvironmental effects,* such as job stress or VDT use. He also suggested that environmental, occupational, and other psychosocial factors may interact to create a total stress load on a worker that is sufficient to precipitate SBS symptoms. Hedge's model represents SBS and other office comfort and health issues as the outcome of the combined influences on the worker of three interlocking subsystems: environmental, building, and work conditions.

In this model, the environmental subsystem includes interrelationships between environmental services (ventilation and lighting systems) and ambient conditions (IAQ, thermal conditions, noise, and vibration). Provision of satisfactory IEQ is influenced by outdoor air quality, standard of HVAC cleaning and maintenance, HVAC design and operation, and the levels of internally generated pollutants from people, materials, and machines. HVAC systems also produce noise and vibrations from both the operations of the plant and the movement of air through the system; such vibrations may contribute to SBS symptoms (Hodgson et al., 1987). HVAC systems function to maintain comfortable thermal conditions, yet by definition even when the ASHRAE thermal comfort standard (ASHRAE, 1992) is met, up to 20 percent of workers may be dissatisfied with thermal conditions. Unsatisfactory thermal conditions may influence SBS symptoms (Jaakkola et al., 1989). Thermal conditions can also affect levels of pollutants by changing emissions of VOCs from many building materials and pressed-wood products, with emissions increasing with increased temperature (Girman et al., 1987; Tichenor, 1987). Many large offices have ceilings finished with sound-absorbing tiles. Acoustic ceiling tiles may be associated with SBS symptoms (Hedge et al., 1993b). Lighting fixtures can affect IAQ, and volatilized asphalt from malfunctioning fluorescent lamp ballasts may be a cause of some SBS symptoms (Tavris et al., 1984). The environmental subsystem can exert direct effects on worker health via poor IAQ, poor lighting, and acoustic and vibration problems, as well as indirect effects, such as workers' negative perceptions of office environmental conditions.

The building subsystem affects the environmental subsystem. Tightly sealing the building shell to conserve energy reduces the rate of natural outdoor air infiltration. The choice of building materials, furnishings, and finishes affects IAQ because these determine what pollutants will off-gas indoors. The layout of office floors, especially the amount and style of office partitioning, affects office ventilation. High partitions can act as barriers to air movement and mixing, creating dead zones in the office. The amount of open shelving (shelf factor) and the area of fleecy-material surfaces (fleece factor) affect SBS symptoms; the higher the shelf and fleece factors, the higher the prevalence of symptoms. The area and type of glazing affects office lighting and thermal conditions. Robertson et al. (1989) reported associations between work-related headache and office-lighting comfort; they suggested maximizing the use of natural light and untinted windows. The building also indirectly affects workers' perceptions of environmental quality, patterns of working, and work-space preferences—for example, many office workers prefer private enclosed offices to open-plan office layouts. The higher levels of distraction in open-plan offices also may be stressful to workers (Hedge, 1986).

The work subsystem influences the activities of office workers; these activities in turn directly affect IEQ and the performance of both the environmental and building subsystems. Breathing generates carbon dioxide, tobacco smoke releases gaseous and particulate air pollutants, carbonless copy paper may release formaldehyde, photocopying may release ozone and some VOCs, and wet photo-copying can be a source of VOCs in buildings (Tsuchiya et al., 1988). Meetings in open office areas can create bothersome noise disturbance for uninvolved workers nearby. Work activities such as computer use repeatedly have been shown to influence SBS symptoms (Hedge et al., 1989b, 1992, 1995, 1996; Skov et al., 1989; Stenberg and Wall, 1995). Inappropriate office lighting can cause re-flected screen glare, existing work furniture may not be ergonomically designed, and the HVAC system may have trouble coping with the additional heat load and possible pollutant load generated by widespread computerization of an office floor. Computer use changes work content as well as the work environment, and together these factors may increase job stress (Hedge et al., 1992), which in turn may precipitate SBS symptoms or change a worker's sensitivity to the ambient environment and work conditions.

Finally, Hedge's model incorporates the idea that in the absence of direct ef-fects of adverse environmental conditions on health, SBS symptoms may be de-pendent on that occupant's psychological state. At any time a person's psycho-logical state results from comparison of the perceived external state of the environment (e.g., presence or absence of a perceived threat) and that person's perceived internal state of his or her body (e.g., presence or absence of a per-ceived symptom). A number of psychobiological factors will affect a person's perceptions of the external and internal conditions (Pennebaker, 1982). Most SBS studies have used self-administered questionnaires to collect symptom data. Self-reports can be influenced by many psychological factors, including a worker's awareness of bodily sensations, and his or her tendency to attribute these to environmental conditions. Pennebaker (1982) eloquently described how physical symptoms are percepts affected by the same cognitive processes that in-fluence all other aspects of perception.

SUMMARY

This chapter summarized research on the causes of SBS. Research results show that symptoms frequently are not associated with prevailing physical environ-mental conditions, apart from sporadic findings of associations between office dust and symptoms. Results have consistently shown associations between SBS symptoms and a variety of psychosocial factors, some of which are occupational, some biological, and some psychological. In its first report on SBS, the World Health Organization pointed to the possible role of psychological factors in in-fluencing SBS complaints. A more detailed consideration of the role of these psychological factors in the etiology of SBS is presented in Chapter 6.

REFERENCES

Armstrong, C.W., Sheretz, P.C., and Llewellyn, G.C. (1989) Sick building syndrome traced to excessive total suspended particulates (TSP), 1AO'89. *The Human Equation: Health and Comfort*, American Society of Heating, Refrigerating, and Air-conditioning Engineers, Atlanta, Georgia, pp. 3–7.

American Society of Heating, Refrigerating and Air Conditioning Engineers (ASHRAE). (1989) *Ventilation for Acceptable Indoor Air Quality, 62–1989*. Atlanta: ASHRAE.

ASHRAE. (1992) *Thermal Environmental Conditions for Human Occupancy, 55–1992R*. Atlanta: ASHRAE.

Burge, P.S., Hedge, A., Wilson, S., Harris-Bass, J., and Robertson, A.S. (1987) Sick building syndrome: A study of 4373 office workers. *Annals of Occupational Hygiene* 31, 493–504.

Burge, P.S., Robertson, A.S., and Hedge, A. (1992) Comparison of self-administered questionnaire with physician diagnosis of the sick building syndrome. *Indoor Air* 6, 422–427.

Burge, P.S., Robertson, A., and Hedge, A. (1993) The development of questionnaire suitable for the surveillance of office buildings to assess the building symptom index, a measure of the sick building syndrome. *Indoor Air '93, Proceedings of the 6th International Conference on Indoor Air Quality and Climate, Helsinki, Finland, July 4–8*. Vol. 1, pp. 731–736.

Eriksson, N., Hoog, J., Stenberg, B., and Sundell, J. (1996) Psychosocial factors and the "Sick Building Syndrome": A case referent study. *Indoor Air* 6, 101–110.

Finnegan, M.J., and Pickering, A. C. (1987) Prevalence of symptoms of the sick building syndrome in buildings without expressed dissatisfaction. In B. Siefert, H. Esdon, N. Fischer, H. Ruden, and J. Wegner (eds.), *Indoor Air '87, Proceedings of the 4th International Conference on Indoor Air Quality and Climate, Berlin (West): Institute for Water, Soil and Air Hygiene*. Vol, 2, pp. 542–546.

Finnegan, M.J., Pickering, A.C., and Burge, P.S. (1984) The sick building syndrome: Prevalence studies. *British Medical Journal* 289, 1573–1575.

Fisk, W.J., Mendell, M.J., Daisey, J.M., Faulkner, U., Hodgson, A.T., Nematollahi, M., and Macher, J.M. (1993) Phase 1 of the California healthy building study: A summary. *Indoor Air* 3 (4) 246–254.

Girman, J., Alevantis, L., Kulasingam, G., Petreas, M., and Webber, L. (1987) The bake-out of an office building: A case study. *Environment International* 15, 449–454.

Godish, T., and Spengler, J.D. (1996) Relationships between ventilation and indoor air quality: A review. *Indoor Air* 6, 135–145.

Gravesen, S., Ipsen, I., and Skov, P. (1993) Partial characterization of the components in the macromolecular organic dust (MOD) fraction and their possible role in the sick building syndrome (SBS). *Proceedings of Indoor Air '93,* Vol. 4, pp. 33–35.

Gyntelberg, F., Suadicani, P., Nielsen, J.W., Skov, P., Valbjørn, O., Nielsen, P.A., Schneider, T., Jorgenson, O., Wolkoff, P., Wilkins, C.K., Gravesen, S., and Norn, S. (1996) Dust and the sick building syndrome. *Indoor Air,* 4 (4), 223–238.

Harrison, J., Pickering, A.C., Finnegan, M.J., and Austwick, P.K.C. (1987) The sick building syndrome: Further prevalence studies and investigation of possible causes. In B. Siefert, H. Esdon, M. Fischer, H. Ruden, and J. Wegner (eds.), *Indoor Air '87: Proceedings of the 4th International Conference on Indoor Air Quality and Climate, Berlin (West): Institute for Water, Soil and Air Hygiene,* Vol, 2, pp. 487–491.

Harrison, J., Pickering, A. C., Faragher, E.B., and Austwick, P.K.C. (1990) An investigation of the relationship between microbial and particulate indoor air pollution and the sick building syndrome. In *Indoor Air '90: Proceedings of the 5th International Conference on Indoor Air Quality and Climate, Toronto, Canada.* Vol. 1, pp. 149–154.

Hedge, A. (1984) Suggestive evidence for a relationship between office design and self-reports of ill-health among office workers in the United Kingdom. *Journal of Architectural and Planning Research* 1, 163–174.

Hedge, A. (1986) Open vs. enclosed workspaces. In J. Wineman, ed., *Behavioral Issues in Office Design.* New York: Van Nostrand Reinhold, pp. 139–176.

Hedge, A. (1988) Job stress, job satisfaction, and work-related illness in offices. In *Proceedings of the 32nd Annual Meeting of the Human Factors Society,* Vol. 2, Santa Monica, CA, pp. 777–779.

Hedge, A. (1990) Sick building syndrome correlates with complex array of factors. *International Facilities Management Journal,* January/February, 52–58.

Hedge, A., Burge, P.S., Wilson, A.S., and Harris-Bass, J. (1989a). Work-related illness in office workers: a proposed model of the sick building syndrome. *Environment International* 15, 143–158.

Hedge, A., Storling, T.D., Sterling, E.M., Collett, C.W., Sterling, D.A., and Nie, V. (1989b) Indoor air quality and health in two office buildings with different ventilation systems. *Environment International* 15, 115–128.

Hedge, A., Erickson, W., and Rubin, G. (1992) Effects of personal and occupational factors on sick building syndrome reports in air conditioned offices. In J.C. Quick, L.R. Murphy, and J.J. Hurrell Jr., eds., *Work and Well-Being: Assessments and Interventions for Occupational Mental Health.* Washington, D.C.: American Psychological Association, pp. 286–298.

Hedge, A., Mitchell, G.E., McCarthy, J., and Ludwig, J. (1993a) Effects of a furniture-integrated breathing-zone filtration system on indoor air quality, sick building syndrome, productivity, and absenteeism. In *Indoor Air '93,*

Proceedings of the 6th International Conference on Indoor Air Quality and Climate, Helsinki, Finland, July 4–8. Vol. 5, pp. 383–388.

Hedge, A., Erickson, W.A., and Rubin G., (1993b) Effects of man-made mineral fibers in settled dust on sick building syndrome in air conditioned offices. In *Indoor Air '93: Health Effects, Proceedings of the 6th International Conference on Indoor Air Quality and Climate, Helsinki, Finland, July 4–8.* Vol. 1, pp. 291–296.

Hedge, A., Erickson, W.A., and Rubin, G. (1994) The effects of alternative smoking policies on indoor air quality in 27 office buildings. *Annals of Occupational Hygiene* 38, 265–278.

Hedge, A., Erickson, W.A., and Rubin, G. (1995) Psychosocial correlates of sick building syndrome. *Indoor Air* 5, 10–21.

Hedge, A., Erickson, W.A., and Rubin, G. (1996) Predicting sick building syndrome at the individual and aggregate levels. *Environmental International* 22, 3–19.

Helsing, K.J., Billings, C.E., and Conde, J. (1987) Cure of a sick building: A case study. In B. Siefert, H. Esdorn, M. Fischer, H. Ruden and J. Wegner, eds., *Indoor Air '87: Proceedings of the 4th Annual Conference on Indoor Air Quality and Climate, Berlin, Germany.* Vol. 3, pp. 557–561.

Hodgson, M.J., and Collopy, P. (1989) Symptoms and the micro environment in the sick building syndrome: A pilot study. In Atlanta, Ga.: ASHRAE, pp. 8–16.

Hodgson, M.J., Permar, E., Squire, G., Cagney, W., Allen, A., and Parkinson, D.K. (1987) Vibration as a cause of "tight building syndrome" symptoms. In B. Siefert, H. Esdorn, M. Fischer, H. Ruden, and J. Wegner eds., *Indoor Air '87: Proceedings of the 4th International Conference on Indoor Air Quality and Climate,* Berlin *(West): Institute for Water, Soil and Air Hygiene.* Vol. 2, pp. 449–453.

Hodgson, M.J., Frohliger, J., Permar, E., Tidwell, C., Traven, N.D., Olenchock, S.A., and Harp, E.M. (1991) Symptoms and microenvironmental measures in nonproblem buildings. *Journal of Occupational Medicine* 33, 527–533.

Hodgson, M.J., Muldoon, S., Collopy, P., and Olesen, B. (1992) Sick building syndrome symptoms, work stress, and environment measures. In *Proceedings of IAQ '92: Environments for People.* Atlanta: ASHRAE, pp. 47–56.

Jaakkola, J.J.K., Heinonen, O.P., and Seppanen, O. (1989) Sick building syndrome, sensation of dryness and thermal comfort in relation to room temperature in an office building: Need for individual control of temperature. *Environment International* 15, 163–168.

Jaakkola, J.J.K., Miettinen, O.S., Komulainen, K., Tuomaala, P., and Seppanen, O. (1990) The effect of air recirculation on symptoms and environmental complaints in office workers: A double-blind, four period, cross-over study. In *Indoor Air '90, Proceedings of the 5th International Conference On Indoor Air Quality and Climate, Toronto, Canada.* Vol. 1, pp. 281–296.

Jaakkola, J.J.K., Heinonen, O.P., and Seppanen, O. (1991) Mechanical ventilation in office buildings and the sick building syndrome: An experimental and epidemiological study. *Indoor Air* 1, 111–121.

Jaakkola, J.J.K., Miettinen, P., Tuomaala, P., and Seppanen, O. (1993) The Helsinki office environment study: The type of ventilation system and the sick building syndrome. In *Indoor Air '93: Health Effects: Proceedings of the 6th International Conference on Indoor Air Quality and Climate, Helsinki, Finland, July 4–8.* Vol. 1, pp. 285–290.

Jaakkola, J.J., Tuamaala P., and Seppanen, O. (1994) Air recirculation and sick building syndrome: Blinded crossover trial. *American Journal of Public Health* 84, 422–428.

Klitzman, S., and Stellman, J.M. (1989) The impact of the physical environment on the psychological well-being of office workers. *Social Science and Medicine* 29, 733–742.

Leinster, P., Raw, G., Thomson, N., Leaman, N., and Whitehead, C. (1990) A modular longitudinal approach to the investigation of sick building syndrome. In *Proceedings of Indoor Air '90, the 5th International Conference on Indoor Air Quality and Climate, Toronto, Canada.* Vol. 1, pp. 287–292.

Lenvik, K. (1990) Comparisons of working conditions and "sick building syndrome" symptoms among employees with different job functions. In *Proceedings of Indoor Air '90, the 5th International Conference on Indoor Air Quality and Climate, Toronto, Canada.* Vol. 1, pp. 507–512.

Mendell, M.J. (1993) Nonspecific symptoms in office workers: A review and summary of the epidemiologic literature. *Indoor Air* 3, 227–236.

Mendell, M.J., and Smith, A.H. (1990) Consistent patterns of elevated symptoms in air-conditioned office buildings: A reanalysis of epidemiologic studies. *American Journal of Public Health* 80, 1193–1199.

Mendell, M.J., Fisk, W.J., Deddens, J.A., Seavey, W.G., Smith, A.H., Smith, D.F., Hodgson, A.T., Daisey, J.M., and Goldman, L.R. (1996) Elevated symptom prevalence associated with ventilation type in office buildings. *Epidemiology* 7, 583–589.

Menzies, R.I., Tamblyn, R.M., Tamblyn, R.T., Farant, J.P., Hanley, J., and Spitzer, W.O. (1990) Sick building syndrome: The effect of changes in ventilation rates on symptom prevalence: the evaluation of a double blind experimental approach. In *Proceedings of Indoor Air '90, the 5th International Conference on Indoor Air Quality and Climate, Toronto, Canada.* Vol. 1, pp. 519–524.

Menzies, R.I., Tamblyn, R.M., Farant, J.P., Hanley, J., Nunes, F., and Tamblyn, R.T. (1993) The effect of varying levels of outdoor-air supply on the symptom of sick building syndrome. *New England Journal of Medicine* 12, 821–827.

Morris, L., and Hawkins, L. (1987) The role of stress in the sick building syndrome. In B. Seifert, H. Esdon, M. Fischer, H. Ruden, and J. Wegner, eds.,

Indoor Air '87, Proceedings of the 4th International Conference on Indoor Air Quality and Climate, Berlin (West), Institute for Water, Soil and Air Hygiene. Vol. 2, pp. 556–571.

Norback, D., Rand, G., Michel, I., and Amcoff, S. (1989) The prevalence of symptoms associated with sick buildings and polluted industrial environments as compared to unexposed reference groups without expressed dissatisfaction. *Environment International* 15, 85–94.

Norback, D., Michel, I., and Widstrom, J. (1990) Indoor air quality and personal factors related to the sick building syndrome. *Scandinavian Journal of Work Environment and Health* 16, 121–128.

Norback, D., Wieslander, G., Strom, G., and Edling, C. (1995) Exposure to volatile organic compounds of microbial origin (MVOC) during indoor application of water-based paints. *Indoor Air* 5, 166–170.

Pennebaker, J.W. (1982) *The Psychology of Physical Symptoms.* New York: Springer-Verlag.

Raw, G.J., Roys, M.S., and Whitehead, C. (1993) Sick building syndrome: Cleanliness is next to healthiness. *Indoor Air* 3, 237–245.

Robertson, A.S., Burge, P.S., Hedge, A., Sims, J., Gill, F.S., Finnegan, M., Pickering, C.A.C., and Dalton, G. (1985) Comparison of health problems related to work and environment measurements in two office buildings with different ventilation systems. *British Medical Journal* 291, 373–376.

Robertson, A.S., McInnes, M., Glass, D., and Burge, P.S. (1989) Building sickness, are symptoms related to office lighting? *Annals of Occupational Hygiene* 33, 47–59.

Skov, P., Valbjørn, O., Pederson, B.V., and Danish Indoor Study Group. (1989) Influence of personal characteristics, job related factors and psychosocial factors on the sick building syndrome. *Scandinavian Journal of Work, Environment and Health* 15, 286–296.

Skov, P., Valbjørn, O., Pederson, B.V., and Danish Indoor Study Group. (1990) Influence of indoor climate on the sick building syndrome in an office environment. *Scandinavian Journal of Work, Environment and Health* 16, 363–371.

Sterling, E., and Sterling, T. (1983) The impact of different ventilation levels and fluorescent lighting types on building illness: An experimental study. *Canadian Journal of Public Health* 74, 385–392.

Stenberg, J., and Wall, S. (1995) Why do women report sick building syndrome symptoms more often than men? *Social Science and Medicine* 40, 491–502.

Strom, G., Norback, D., West, J., Wessen, B., and Palmgren, U. (1993) Microbial volatile organic compounds (MVOC)—a causative agent to sick building problems. In E. Sterling, C. Bieva, and C. Collett, eds., *Building Design, Technology and Occupant Well-being in Temperate Climates, Proceedings of the International Conference, ASHRAE, Brussels, Belgium, February 17–19.* pp. 351–357.

Sundell, J., Andersson B., Andersson K., and Lindvall, T. (1993) Volatile organic compounds in ventilating air in buildings at different sampling points in the buildings and their relationship with the prevalence of occupant symptoms. *Indoor Air* 2, 82–93.

Sundell, J. (1994) On the association between building ventilation characteristics, some indoor environmental exposures, some allergic manifestations and subjective symptom reports. *Indoor Air* 2 (Supplement), 1–42.

Tamblyn, R.M. and Menzies, R.I. (1992) Big sufferers of work-related symptoms in office buildings—who are they? *IAQ '92: Environments for People*, Atlanta: ASHRAE, pp. 300–308.

Tamblyn, R.M., Menzies, R.I., Nunes, F., Leduc, J., Pasztor, J., and Tamblyn, R.T. (1993) Big air quality complainers—are their office environments different from workers with no complaints? In *Indoor Air '93, Proceedings of the 6th International Conference on Indoor Air Quality and Climate, Helsinki Finland, July 4–8*. Vol. 1, pp. 133–140.

Tavris, D.R., Field, L., and Brumback, C.L. (1984) Outbreak of illness due to volatilized asphalt coming from a malfunctioning fluorescent lighting fixture. *American Journal of Public Health* 74, 614–615.

Teeuw, K.B., Vandenbroucke-Grauls, C.M.J.E., and Verhoef, J. (1994) Airborne gram-negative bacteria and endotoxin in sick building syndrome. *Archives of Internal Medicine* 154, 2339–2345.

Tichenor, B.A. (1987) Organic emissions via small chamber testing. In B. Seifert, H. Esdon, N. Fischer, H. Ruden, and J. Wegner, eds., *Indoor Air '87, Proceedings of the 4th International Conference on Indoor Air Quality and Climate, Berlin (West): Institute for Water, Soil and Air Hygiene*. Vol. 1, pp. 8–13.

Tsuchiya, Y., Clermont, M.J., and Walkinshaw, D.S. (1988) Wet process copying machines: A source of volatile organic compound emissions in buildings. *Environmental Toxicology and Chemistry* 7, 15–18.

Valbjørn, O., and Skov, P. (1987) Influence of indoor climate on the sick building syndrome prevalence. In B. Seifert, H. Esdon, N. Fischer, H. Ruden, and J. Wegner, eds., *Indoor Air '87, Proceedings of the 4th International Conference on Indoor Air Quality and Climate, Berlin (West): Institute for Water, Soil and Air Hygiene*. Vol. 2, pp. 593–597.

Wallace, L.A., Nelson, C.J., and Dunteman G. (1991) Workplace characteristics associated with health and comfort concerns in three office buildings in Washington D.C. In *Proceedings of IAQ '91: Healthy Buildings*. Atlanta, Ga.: ASHRAE, pp. 56–60.

Wilson, S., and Hedge, A. (1987) *The Office Environment Survey: A Study of Building Sickness*. London: Building Use Studies.

Wilson S., O'Sullivan P., Jones P., and Hedge A. (1987) *Sick Building Syndrome and Environmental Conditions: Case Studies of Nine Buildings*. London: Building Use Studies.

World Health Organization (WHO). (1983) Indoor air pollutants: Exposure and health effects. *EURO Reports and Studies* 78.

Wyon, D. (1992) Sick buildings and the experimental approach. *Environmental Technology* 13, 313–322.

Zweers T., Preller L., Brunekreef B., and Bloeij, J.S.M. (1990) Relationships between health and indoor climate complaints and building, workplace, job and personal characteristics. In *Indoor Air '90: Proceedings of the 5th International Conference on Indoor Air Quality and Climate, Toronto, Canada.* Vol. 1, pp. 495–500.

Zweers, T., Preller, L., Brunekreef, B., and Boleij, J.S.M. (1992) Health and indoor climate complaints of 7,043 office workers in 61 buildings in the Netherlands. *Indoor Air* 2, 127–136.

Chapter 6

Investigating Health Complaints: Behavioral Aspects

ALAN HEDGE, PH.D.

This chapter examines the behavioral factors that have been shown to influence the perception of indoor quality and occupant reports of indoor environmental quality (IEQ), sometimes called sick building syndrome (SBS), in offices.

BUILDING STANDARDS

When Tredgold first formulated building-ventilation standards more than 150 years ago, he specified ventilation rates designed to minimize problems of odors and stale air in buildings. This fundamental goal has remained unchanged. The ventilation standard should specify ventilation system design that (1) ensures that the ambient air conditions satisfy human requirements for fresh air that is free from malodors and pollutants at hazardous levels, and (2) creates conditions of thermal comfort for building occupants.

The American Society of Heating, Refrigerating, and Air Conditioning Engineers (ASHRAE) Standard 62-1989R defined acceptable indoor air quality (IAQ) as air that is free from contaminants at hazardous concentrations, as determined by cognizant authorities, and with which a substantial majority of occupants, 80 percent or more, report satisfaction (ASHRAE, 1989). Thus, the creation of a satisfactory indoor climate is inextricably connected to human perception of environmental conditions. Moreover, given the ASHRAE 62-1989R definition of acceptable indoor air quality, a determination of whether or

not a building satisfies ASHRAE requirements cannot be made without knowledge of the occupants' experiences and opinions.

However, research studies have shown that complaints of poor IEQ and SBS
can be prevalent even in buildings that meet the current ASHRAE standard. Consequently, before concluding that a genuine ventilation problem exists in any
building, it is necessary to examine the roles of various nonenvironmental factors that can influence human complaints and can frequently mislead investigators attempting to diagnose genuine building-ventilation problems.

SICK BUILDINGS OR SICK PEOPLE?

The term *sick building* is evocative and much-beloved by the media. The term
has been widely and indiscriminately applied to buildings in which ventilation
problems are suspected. However, it is worth remembering that it is the occupants inside the building who report symptoms, and not the building itself that
gets sick. Buildings without occupant complaints are often described as healthy
buildings, while those with a high prevalence of complaints are termed sick
buildings. However, the presence or absence of complaints alone cannot be re-
lied on for an accurate diagnosis of conditions within a building, because the
presence of many atmospheric hazards, such as microorganisms (viruses, bacteria), carbon monoxide and dioxide, radon, among others, usually is undetectable
to the occupant until the body has reacted in some way, such as when an exposed
person reports a clinical symptom and shows a physical sign of illness.

The World Health Organization (WHO) (1983) defined SBS as a collection
of self-reported symptoms of general malaise, including eye, nose, throat, and
skin irritation; headache; lethargy; and difficulty breathing. WHO recognized
that each of these symptoms is relatively common among the general population
but that together this constellation of symptoms could be much more prevalent
among occupants in certain buildings. Occupants who frequently experience
these SBS symptoms while in the building typically report that their symptoms
lessen when they leave the building. Clinical signs of illness, such as fever, invariably are absent in these occupants, complicating a medical diagnosis.

Thus, IEQ, or SBS is not a building-related illness, such as Legionnaires'
disease, and the pattern of symptom reporting does not follow that of any known
diseases. Wisely, WHO also noted that psychological processes might mediate
reports of IEQ, although they may not be the main cause of this syndrome. This
observation seems to have been overlooked by all but the few researchers studying this phenomenon.

Over the past decade, a considerable body of research evidence has been ac-
cumulated that shows self-reported IEQ symptoms correlate seldom with indoor

air contaminants but usually with many nonenvironmental variables that influence our perception of health.

All of the published research studies on IEQ reviewed for this chapter found that IEQ symptoms are reported by some workers in every building investigated. Indeed, it is probably true that in all large buildings, some occupants will report IEQ problems that they attribute to the building's air quality. Thus, if IEQ reports alone are used as the criterion for judging the performance of a building, all buildings will fall on a continuum between low and high levels of IEQ reports, and choosing a criterion for categorizing a building as "sick" or "healthy" will be purely arbitrary.

Also, the prevalence of symptom reporting may fluctuate or remain relatively stable within the same building irrespective of any variations in the physical environmental conditions. In a so-called sick building, a majority of the occupants report ill health that they attribute to their building, but this does not necessarily mean that these symptoms arise solely because IAQ is compromised. Menzies et al. (1994) reported a study of 2,317 workers from 11 "nonproblem" and five "problem" office buildings in which they found no significant differences in the prevalence of IEQ symptoms between buildings that had been labeled as either sick or healthy by their management, and no associations between symptoms and measured environmental conditions. Indeed, there was a complete overlap in the prevalence of symptoms between buildings. The researchers concluded that "to label an entire building as sick or healthy has no scientific foundation" (p. 37). However, others have suggested that "a number of different pollutant categories clearly contribute to such symptoms" (Apter et al., 1994).

RESULTS FROM IEQ RESEARCH STUDIES

The conventional and widely held view of SBS is that this is a disorder caused primarily by inadequate ventilation and consequent poor IAQ. This view suggests that a lack of outdoor air in a building leads to the accumulation of indoor pollutants and that these pollutants are responsible for IEQ symptoms among occupants. It follows from this view that increasing the outdoor air supply to buildings should reduce the prevalence of IEQ symptoms and that IEQ symptoms should correlate with the concentrations of one or more indoor pollutants. In spite of more than a decade of research, neither of these sequelae has been established.

Several studies investigated the effects of increasing the ventilation rate on IAQ and IEQ complaints; results have failed to show any reduction in IEQ symptoms that parallels changes in ventilation rates (e.g., Menzies et al., 1993; Nagda et al., 1991). Similarly, epidemiological studies have been undertaken in the United States (Hedge et al., 1992, 1993b, 1995, 1996), in Canada (Menzies et al., 1993), in the United Kingdom (Burge et al., 1987; Finnegan et al., 1984; Harrison et al., 1990), in Sweden (Nörback et al., 1993; Stenberg and Wall, 1995), in Finland (Jaakkola et al., 1991), in Denmark (Skov et al., 1987, 1989), and in the Netherlands (Zweers et al., 1992). In spite of sporadic associations, none of these studies established any consistent correlation between IEQ symptoms and indoor pollutants, although each of these studies found that nonenvironmental factors consistently are associated with IEQ symptoms.

Research studies repeatedly have shown that SBS symptoms are more prevalent among workers in air-conditioned offices than in naturally ventilated offices (Chandrakumar et al., 1994; and see Mendell, 1993, and Sundell, 1994, for reviews). Only occasionally have IEQ studies found indoor contaminants that correlate with IEQ symptom prevalence, and these contaminants have been components of office dust (e.g., Hedge et al., 1993; Gyntelberg et al., 1994). Leinster et al. (1990) and Raw et al. (1993) found that the thorough cleaning of offices can dramatically reduce IEQ symptoms.

Usually, research studies found that IEQ reports correlate with the following diverse range of nonenvironmental variables:

- Worker's gender—women report more IEQ symptoms than men in the same environment
- Computer use—symptom reports increase with daily computer use over 1 hour
- Job stress—symptom reports increase with stressful work demands
- Job satisfaction—symptom reports increase with increasing dissatisfaction with work
- Atopy—symptom reports are higher among those with respiratory allergies
- Migraine—symptom reports are higher among migraine headache sufferers
- Eyewear—symptom reports are higher among workers who wear corrective lenses (eyeglasses, contact lenses)
- Age—symptom reports are higher among workers under age 35 years than over age 35
- Smoking—symptom reports are slightly higher for smokers

Research studies repeatedly have shown that SBS symptoms are more prevalent among workers in air-conditioned offices than in naturally ventilated offices; most of these studies also found stronger associations between a variety of organizational, occupational, and personal variables and symptoms than between these and environmental conditions (Chandrakumar et al., 1994; and see Mendell, 1993, and Sundell, 1994, for reviews).

Studies of office workers also have shown that although workers believed their symptoms were caused by prevailing environmental conditions, there was no correlation between these conditions, such as daily variations in temperature and humidity, and IEQ symptoms, such as dry nose and nasal congestion complaints, whereas IEQ symptoms were correlated with stress levels (see, e.g., Morris and Hawkins, 1987).

The prevalence of IEQ symptoms among office workers is influenced by occupational factors (job level, hours of computer use, job stress, job satisfaction, handling of carbonless copy paper, photocopying), psychological factors (perceptions of control, perceptions of ambient conditions, perceptions of comfort), personal factors (gender), and organizational factors (public-sector vs. private-sector buildings) (Hedge, 1988; Hedge et al., 1989; Skov et al., 1989). In a study of almost 4,500 office workers in air-conditioned office buildings, perceived indoor quality, computer use, gender, job satisfaction, and job stress significantly influenced the number of IEQ symptoms reported by workers (Hedge et al., 1992, 1993b, 1995; Tamblyn et al., 1993). Thus, research findings call into question the direct role of environmental variables alone in the etiology of IEQ.

PERSONAL INFLUENCES ON REPORTING IEQ

Perceptions of Indoor Air Quality

Complaints of poor IAQ and IEQ usually are perplexing to diagnose, because results of IAQ surveys often fail to find contaminated air, and simply increasing the ventilation rate often fails to resolve IAQ problems. Even back in 1983, WHO recognized that IAQ complaints and IEQ symptoms are influenced by various nonenvironmental variables such as personal, occupational, and psychological factors. These factors may act to modify a person's susceptibility and sensitivity to IAQ problems. So understanding the role these variables can play may assist in resolving IAQ issues.

Perception of IAQ is at the root of all judgments that we make about the potential hazards in the air we breathe. Perception of IAQ depends on various sensory processes. Biologically speaking, lay people are quite ill-equipped to make judgments about many indoor air contaminants.

We do not, for example, possess adequate sensory apparatus to detect any of the following common airborne contaminants and conditions at the levels typically found inside buildings:

- Carbon monoxide
- Carbon dioxide
- Radon, thoron
- Asbestos, mineral fibers

- Viruses, bacteria
- Fungi
- Pollen
- Dust mites, mite feces
- Respirable particulates 10 microns or less in diameter (PM10)
- Nonaromatic volatile organic compounds (VOCs)
- Relative humidity

Airborne contaminants and conditions that people can detect when concentrations are high enough, along with the telltale sensation, include the following:

- Formaldehyde (irritation)
- Aromatic VOCs (odors, irritation)
- Ozone (irritation)
- Air temperature (cutaneous sense)
- Environmental tobacco smoke (odors, irritation, sight)

When an airborne contaminant that can cause mucous membrane irritation, such as formaldehyde, arrives at a mucous membrane, it is detected by the body's general chemical sense. The resulting sensations we experience in the presence of such pollutants are eye, nose, and throat irritation, and we may interpret these sensations as indicators of poor IAQ.

Other airborne molecules can be detected as either pleasant or unpleasant odors. Cain and Cometto-Muñiz (1993) reported that odor thresholds can be predicted from the carbon chain length of the molecule, and that pungency can be predicted from odor thresholds. Odors are detected by specific receptors in the olfactory nasal mucosa, and odors also can signal poor air quality, although perceptions of IAQ based on odor reports often do not correspond to the presence of actual hazards.

For example, the smell of garlic in indoor air may be highly pleasant to one person, who then will judge the air quality as good, whereas another person may find that smell highly unpleasant and will therefore judge the same air quality as bad. People's preferences change over time. For example, a smoker may consider the smell of environmental tobacco smoke desirable but upon quitting smoking may eventually judge the same smell as undesirable. Consequently, sensitivity to irritants and judgments of odor usually are the outcomes of relative but not absolute perceptual processes. Use of these measures as sole criteria to evaluate IAQ is foolish.

Signal-Detection Theory

The sensory mechanisms for detecting the presence or absence of pollutants also behave as a relative system. In sensory psychology, *signal-detection theory* best describes how human sensory processes operate. This theory proposes that an individual's perception of signals, such as odors or irritants, depends on the intensity of the stimulus (termed the *signal*) relative to both background activity in the nervous system and/or irrelevant background competing stimuli (termed *noise*). In other words, what we can and cannot detect depends in part on the signal/noise ratio. For example, in a room full of sweaty clothes, the addition of another item of sweaty clothing probably will not have much effect on perceived IAQ, but in a room full of clean clothes, the addition of the same sweaty item will have a marked detrimental effect on perceived IAQ.

From this simple observation, it is clear that the drive toward creating odor-free buildings is doomed to failure, because as we reduce and remove all odors from indoor air, we effectively enhance the sensitivity of people to the presence of smaller and smaller contributors of any odor in buildings—that is, by reducing the background noise in the building, we make it easier to detect a signal of which we otherwise would have been unaware.

Signal-detection theory postulates four states that can arise in any sensory situation:

- *Hit*—correct identification of a signal when this is present, for example, you smell cigarette smoke when someone is smoking
- *Correct rejection*—correct identification of a nonsignal when the signal is absent, for example, you do not smell smoke and no one is smoking
- *Miss*—failure to identify the presence of a signal when the signal is present, for example, you do not smell smoke but it is present and someone is smoking
- *False alarm*—failure to identify a nonsignal when the signal is absent, for example, you think you smell smoke when there is none and no one is smoking

In any sensory situation we want to try to maximize the hits and correct rejections and minimize the misses and false alarms. Various factors influence how successfully we can do this. First, if the magnitude of the difference between the signal and the background noise (termed d prime [d'] in the theory) is small, we are more likely to make erroneous judgments; whereas if it is large, then we are less likely to make errors. For example, in a smoky room a person is less likely to detect when another cigarette has been lit than when in a smoke-free room. Second, the criterion (this is termed beta [β] in the theory) we use as a basis for deciding between the presence (hit or false alarm) or absence (correct rejection or miss) of a signal depends on the likely outcome of either detecting or failing to detect a significant event.

For example, if I believe that I am especially sensitive to a fragrance and that exposure to this fragrance will be life-threatening, then I am more likely to increase my false alarm rate (i.e., suspect the fragrance is present in many situations where it isn't) than if I have no particular feelings about the fragrance. Signal-detection theory provides a useful framework within which we can begin to understand why sensitivity to irritation and odors is not fixed and why human reactions to perceived indoor air contaminants varies between people and changes over time.

Staudenmeyer et al. (1993) used a signal-detection theory approach to test the effects of odorous air pollutants on 20 individuals presenting with multiple chemical sensitivities (MCS) in a double-blind provocation chamber challenge, using a masking odor for both sham controls and pollutants. Results failed to support a pattern of true chemical sensitivity and suggested that subjects were responding randomly to the actual challenges. MCS was addressed in more detail in Chapter 4.

Other sensory physiological processes, such as adaptation effects, also affect our ability to judge IAQ. For all individuals, perceived intensity of an odor that is emitted at a constant rate varies with time because of sensory adaptation processes, and consequently odor judgments made immediately upon entering a room are more intense than odor judgments made 30 minutes later.

For example, when you first walk into a sweaty locker room you will notice the body odor, but after 15 minutes in the room the body odor will seem much less intense, even though objectively the number of odorous molecules has not decreased. Immediate odor judgments have been suggested as a metric for determining perceived IAQ in buildings (European Concerted Action, 1992), although judgments of what are desirable and undesirable odors are influenced by personal preferences and beliefs. A large-scale study of more than 50 buildings in Europe found no correlations between ratings of indoor odors (expressed as olf units [1]) and either the objective quality of the indoor air, occupant perceptions of IAQ, or IEQ symptoms (Bluyssen et al., 1996).

Even when we can accurately detect what we judge to be unpleasant odors, these do not necessarily indicate hazardous indoor air. Conversely, the absence of odors does not necessarily signify healthy indoor air, because, as we have seen, many pollutants, like carbon monoxide, carbon dioxide, and airborne microorganisms, have no odor and cannot be detected by our sensory systems. The extent to which odors may present any significant health risk to individuals remains debatable. Research on human exposures to a VOC mixture suggests that odor complaints may precede irritation phenomena and that as odor adapta-

---

[1] Fanger (1988, 1989) developed two new units of odor measurement—the olf, which quantifies pollutant emissions from sources, and the decipol, which quantifies airborne pollutant concentrations. By definition, 1 olf is the emission rate of air pollutants (bioeffluents) from a standard person (1.8 square meters body area, 0.7 baths per week, clean underwear daily, and 80 percent wearing deodorant); the decipol is the concentration of air pollutants from a 1-olf source ventilated by 10 liters/second (approximately 20 cubic feet per minute) of unpolluted air.

tion occurs, irritation complaints may increase (Otto et al., 1993). Also, evidence exists of a gender difference in odor sensitivity. Women performed better on a recognition memory test for 40 odors than did men (Lehrner, 1993). Gender differences have also appeared for odor intensity and hedonic ratings (Doty et al., 1984) and odor identification (Cowart, 1989; Doty et al. 1985; Wysocki and Gilbert, 1989).

In addition to the possible roles of irritation and odors in affecting judgments of IAQ, field research shows that thermal comfort variables, such as air temperature and air movement, affect perceptions of stale and stagnant air. Experiments show that judgments of acceptability correlate better with the percentage of dissatisfied people than judgments of odor intensity. Acceptability ratings change with air temperature (from 68°F [20°C] to 79°F [26°C]), and the higher the air temperature, the lower the acceptability of perceived air quality (Iwashita, 1992).

Recently, attention has focused on the possibilities of regulating chemical emissions from certain indoor products, especially those with odorous emissions such as carpet and scented products, because of fears that these emissions might be linked with MCS and SBS, even though no scientific evidence exists to support this view. It is therefore sensible to explore the prevalence of concerns about odors and chemical sensitivities among the normal occupants of buildings.

Cacosmia

In a 1993 study, *cacosmia* was defined as an altered sense of smell, accompanied by a tendency to feel ill (e.g., nausea, headache, dizziness) from the odor of chemicals at low levels that have no effect on normal individuals (Bell et al., 1993). This term is not a clinical definition (i.e., there exists no objective evidence of altered odor sensitivity) but rather a self-reported definition.

In their study, Bell et al. asked people five questions; based on the responses, people were categorized as either cacosmic or noncacosmic. Responses to sensitivity questions were used to create a multiple chemical sensitivity category and a cacosmia category (Bell et al., 1993). To be considered multiply chemically sensitive, a person had to answer in the affirmative to any three of the four change item questions (sensitivity to foods; sensitivity to home or home furnishings; need to wear special clothing because of chemical sensitivity; have trouble shopping in stores or eating in restaurants because of chemical sensitivity).

To be considered cacosmic in the Bell study, a person had to respond that any three or more of the five chemical sensitivity criteria (paint, perfume, pesticides, carpet, car exhaust) caused illness sometimes, often, or always. Although only 0.8 percent of respondents fit these criteria for multiple chemical sensitivity, 26.3 percent of office workers surveyed considered themselves "chemically sensitive." About one-third of office workers (36.1 percent) fit the criteria for cacosmia, and interestingly, 22.4 percent of these were current smokers (Hedge and Erickson, 1996).

Thus, even though there is a lack of clinical evidence to support heightened odor and chemical sensitivities among office workers, around one-quarter of the workforce believe that they have such sensitivities.

COGNITIVE INFLUENCES ON PERCEIVED SBS

Contagion Effects

Because we cannot directly sense many IAQ hazards, including airborne bacteria and colorless, odorless, and toxic gases, we must rely on beliefs and imagination to help us anticipate and avoid them. These processes also change how we interpret internal bodily sensations. Belief and imagination processes work to influence what we create or choose as hypotheses to explain what we believe is happening in our environment and inside our bodies.

If events transpire that convince a worker that he or she has been breathing air containing a colorless, odorless, yet noxious pollutant, then that person may begin to manifest the symptoms that they are told are indicators of exposure. For example, a person told that the pollutant causes eye irritation will naturally and subconsciously begin to selectively attend to eye sensations for confirmation of any exposure. The person may even behave in a way that creates this information, such as rubbing the eyes more frequently than normal, which increases irritation sensations, which in turn is interpreted as a sign of exposure rather than merely overrubbing. Such behaviors normally are quite common.

Someone who thinks about an itch on the nose eventually will scratch the nose, even though nothing external was causing the itching sensation. If you think about mites and fleas crawling over your body, you will eventually experience sensations of itchy skin and want to scratch yourself. At musical concerts, audience members are likely to feel the urge to cough when they hear others coughing—hence the outbursts of coughing during intermissions. At comedy shows, audience members are likely to start laughing when they hear others laugh—hence the use of canned laughter on television sitcoms. These behavioral contagion effects are very powerful, and they include the development of imaginary symptoms. Studies of medical students have shown that 70 percent of freshmen believe they have developed the symptoms of diseases being studied (Pennebaker, 1982). Typically, in cases of IEQ, contagion effects have not been systematically investigated. However, such effects have been suspected in at least one major incident involving evacuation of a building (Hedge et al., 1987).

Mistaken Perceptions

Mistaken reporting of bodily sensations is common. Only about half the people who have a balloon inflated in their stomachs report a sensation of feeling full (Pennebaker, 1982). Similar mistakes in interpreting sensations also affect our perception of indoor climate conditions. For example, complaints of sensations of dry air are not good indicators of low relative humidity in the range of 15 percent to 35 percent.

Research shows no significant correlation between measured relative humidity and sensations of dry air in this range for men. Although there is a significant negative correlation between measured relative humidity and perceptions of dry air in this range for women, reports of dryness sensations actually increase as relative humidity increases, the very reverse of expectations (Göthe et al., 1987).

Attentional Focus Effects and Suggestion

Attentional influences affect what information we choose to attend to in our environment and also what symptoms we choose to monitor. IEQ symptoms are percepts that are affected by the same cognitive processes that influence all other aspects of perception (Pennebaker, 1982). When people are told to focus attention on the amount of nasal congestion, increases in reports of nasal stuffiness typically occur; when told to focus attention on free breathing through the nose, decreases in reports of nasal congestion occur under precisely the same environmental conditions (Pennebaker and Skelton, 1981).

Consequently, the use of poorly designed questionnaires with leading, biased, and suggestive questions will inadvertently overestimate the real prevalence of problems in a building. Such overestimation can also occur if the questionnaire is distributed at a time of heightened concern among workers that an invisible hazard may be present in their building.

The power of suggestion is well illustrated by the results of a study conducted by Knasko et al. (1990), who exposed 90 individuals between ages 18 and 35 to a placebo of sprayed water following the suggestion of a pleasant, neutral, or bad odor. The suggestions showed no effect on performance of a clerical coding task; however, mood was more positive for those who experienced the pleasant suggestion, and reports of physical symptoms were higher with the bad odor suggestion.

Knasko (1992) exposed subjects to lemon, lavender, or dimethyl sulfide odors. No differential effects of odors on performance of a creativity task were found. Mood ratings were worse for exposure to the malodor (dimethyl sulfide), but exposure to pleasant odors did not affect mood ratings. Ambient lemon odor decreased the number of perceived health symptoms. Suggestive evidence for a positive relationship between odor effects and personality (external locus of control) was also found.

Conditioned Learning Effect

Occupant reactions to buildings can be affected by conditioned learning. Classical conditioning, first demonstrated by Pavlov, occurs when an unconditioned stimulus (e.g., a bell) occurs contiguously with a conditioned stimulus (e.g., food) that elicits a conditioned response (e.g., salivation). After repeated pairings, presentation of the unconditioned stimulus alone is sufficient to elicit the conditioned response. In some circumstances, classical conditioning can occur very rapidly. Even somatic responses, such as the reaction of the immune system to a challenge, can be classically conditioned (Dantzer, 1993). A person who fears that he or she has been exposed to toxic chemicals will show natural physiological changes, such as increased heart rate, labored breathing, dizziness, and nausea, that reflect increased anxiety. These symptoms can become conditioned to the appropriate stimulus, such as an odor or the sight of a product suspected of toxic emissions (Bolla-Wilson et al., 1988).

Thus, classical conditioning may inadvertently occur in many IAQ situations. Consider, for example, the possibility of conditioning occurring when a person is exposed to a carpet with a new-carpet odor. If we first prepare the person by telling him or her that new-carpet odor is toxic, then exposure to the carpet will pair two sets of stimuli. Both the visual stimulus of the carpet and the presence of an odor will be sufficient to raise the person's anxiety level, which in turn may induce feelings of nausea, dizziness, and breathing difficulty. These physiological responses, the result of anxiety, will then become conditioned to the sight and smell of a new carpet. Such processes can also result in people becoming conditioned against being in a building that they think poses a threat. This helps to explain why some workers ultimately have to quit working in their building, even in the absence of any documented hazard.

Reinforced learning occurs when an additional type of stimulus is added to the learning situation following the behavioral response. When a reward is given, future similar behaviors are encouraged; if a reward is withheld or a punishment is administered, future similar behaviors are discouraged. In a building with suspected air-quality problems, workers may reinforce each others' beliefs about the building by consensually agreeing that a problem must exist. Workers reporting the most severe symptoms may be selected for special treatment, such as relocating to a special office, making special work arrangements, receiving media attention, and the like. Such actions can inadvertently reward a worker for behaving as an ill person and can reinforce a worker's beliefs about his or her exposure and/or illness.

Reinforced learning is also affected by social groups. Some people, such as MCS patients, may have found a complete alternative social-support system that exists to reinforce the person's beliefs in a nonexistent physical illness. Clearly, how management reacts to and handles a suspected IAQ problem in a building will have a major influence on the degree to which these conditioning processes affect occupants.

Attribution Effects

Because of limited biological sensory systems, occupants frequently cannot accurately identify what is causing their symptoms. Headaches, for example, can be caused by poor IAQ, inadequate lighting, pressure of work, noisy working conditions, and so on. Building occupants therefore must rely on personal beliefs about what is causing the sensations they are experiencing. These beliefs will encourage them to attribute a cause to whatever is presented to them as the most salient factor at the time. For example, after a television special about the potential hazards of chloroform fumes from chlorinated water systems has been aired, a building manager can expect an increase in complaints of chlorine odors and associated health problems among those who saw the program. How management interacts with the news media and how the media report IAQ cases have substantial effects on subsequent reports of problems in a building.

PSYCHOSOCIAL INFLUENCES ON REPORTING SBS

Influence of Stress

Studies of IEQ in offices frequently found that work stress plays a significant role in the occurrence of symptoms. The extent to which other stressors, such as those in the physical environment (e.g., poor lighting, poor acoustics, lack of privacy, poor layout, poor ergonomics, work pressure) create strain and adverse health effects may depend on the characteristics of exposed individuals and their ability to cope with these stressors (Hedge et al., 1989). However, evidence on the effects of different coping styles presently is sparse in the literature. IEQ complaints and SBS symptom reports may arise from the combined effects of diverse environmental and nonenvironmental factors, which in turn may stress the body and, depending on individual coping abilities, increase the personal strain that an individual experiences as a series of symptoms. It may be that an increase in personal strain can alter an individual's sensitivity to environmental stimuli such as airborne irritants, or perhaps it might directly cause some of the IEQ symptoms, such as headache, nausea, dizziness, breathing difficulty, and so on. Research on office workers has shown that although workers believed their symptoms were caused by prevailing environmental conditions, there was no correlation between these conditions, such as daily variations in temperature and humidity, and IEQ symptoms, such as dry nose and nasal congestion complaints. However, IEQ symptoms were correlated with stress levels (Morris and Hawkins, 1987). Occupational factors influence the prevalence of IEQ symptoms among office workers; these factors include job level, hours of computer use, job stress, job satisfaction, handling of carbonless copy paper, and photocopying. Psychological factors (perceptions of control, perceptions of ambient conditions, perceptions of comfort), personal factors (gender), and organizational factors (public-sector vs. private-sector buildings) also influence symptoms (Hedge, 1988; Hedge et al., 1989; Skov et al., 1989). In a study of almost 4,500 office workers in air-conditioned office buildings perceived to have indoor quality issues, amount of computer use, gender, job satisfaction, and job stress were shown to significantly influence the number of IEQ symptoms workers reported (Hedge et al., 1992, 1993b; Tamblyn et al., 1993).

Mass Psychogenic Illness

Mass psychogenic illness (MPI) refers to "the collective occurrence of a set of physical symptoms and related beliefs among two or more individuals in the absence of any identifiable pathogen" (Colligan and Murphy, 1982). Social psychological processes of *contagion,* where complaints and symptoms spread from person to person, and *convergence,* where groups of people develop similar symptoms at about the same time, underlie MPI. Environmental events, like an unpleasant odor, can trigger contagion and convergence processes, and occupants who cannot readily identify what has triggered their symptoms often attribute these to any visible environmental changes, such as installation of a new carpet, or invisible agents, such as a "mystery bug." MPI symptoms include headache, nausea, weakness, dizziness, sleepiness, hyperventilation, fainting, vomiting, and occasionally skin disorders and burning sensations in the throat and eyes (Boxer, 1985, 1990; Colligan and Murphy, 1982; Olkinuora, 1984). No credible scientific evidence exists that the odor of new carpets is potentially toxic or that carpet odor, per se, is the trigger for MPI.

MPI reactions probably arise from the interaction of preexisting poor physical environmental conditions (poor ventilation, poor lighting, excessive noise), stressful work conditions (tedious work, poor organizational climate, poor labor-management relations), and differences in disposition among individuals (gender differences, differences in anxiety levels) with a triggering event (bad odor), followed by inappropriate management response to the perceived threat. Studies of MPI typically find a similar sequence of events leading to the incident.

MULTIPLE CHEMICAL SENSITIVITY AND ENVIRONMENTAL ILLNESS

People diagnosed by clinical ecologists as suffering from MCS or environmental illness are claimed to be hypersusceptible to environmental agents (Miller and Ashford, 1993; Rea, 1992; Ziem, 1992). Other terms that have been used to describe this condition include ecological illness, cerebral allergy, total allergy syndrome, and twentieth-century disease. Superficially, the symptoms appear to be similar to those of SBS, that is, they represent a nonspecific malaise. But they also include non-SBS symptoms, such as more-frequent complaints of nausea, breathing difficulties, and gastrointestinal upset. Also, unlike SBS symptoms, the symptoms of MCS are thought to result from specific provocative chemical agents, though these are usually not identified.

MCS symptoms have been subjectively linked to exposure to a variety of modern products, such as new carpet, new clothing, cleaning products, perfumes, cosmetics, mothballs, laser printers, vehicle exhaust, paper, new particleboard and plywood, gas-stove emissions, and pesticides (Hileman, 1991). For example, some 40 to 50 workers claimed to have developed MCS following the installation of new carpeting in the Waterside Mall headquarters of the U.S. Environmental Protection Agency in Washington, D.C. (Hileman, 1991).

The etiology of MCS includes a hypothetical process, termed *limbic kindling,* in which the activities of the olfactory bulb, hypothalamus, and limbic system supposedly are coordinated in such a way that exposure to a minute dose of a sensitizing substance elicits a disproportionate neural response (Miller and Ashford, 1993). Another proposed hypothesis is that MCS is a form of neurogenic inflammation mediated via a neural peptide called substance P (Meggs, 1993). To date, neither of these suggestions is supported by empirical research.

The symptoms reported by many MCS sufferers appear comparable to those of one or more commonly recognized psychiatric disorders, such as mood disorders, affective disorders, and anxiety disorders (Black, Rathe, and Goldstein, 1990). Immunologic abnormality does not seem to be a component of MCS (Hall, 1989). Terr (1986) studied 50 patients with a clinical ecology diagnosis of environmentally induced illness. Immunological measures were unremarkable, including serum levels of immunoglobulins and complement and circulating lymphocyte, B-cell, T-cell, and T-cell subset counts. Patients had symptoms of psychosomatic illness.

Among plastics workers in an aerospace-equipment plant who were known to be exposed to phenol, formaldehyde, and methyl ethyl ketone, comparison of the psychiatric profiles of workers who did and those that did not report chemical sensitivity showed that psychiatric factors predating occupational exposures, especially prior major depression and/or panic disorder, were the stronger predictors of reported chemical sensitivity than actual chemical exposures (Simon et al., 1990). Subsequently, Simon et al. (1993) compared immune function, psychological status, and cognitive function in patients diagnosed by clinical ecologists as suffering from MCS, with normal controls. The immunologic testing included several measures, such as autoantibody titers, lymphocyte surface markers, and IL-1 generation by monocytes. No differences between MCS cases and controls were found for these immunologic measures. Standard psychological measures of anxiety, depression, and somatization showed that MCS sufferers reported more anxiety or depressive disorders, which did not appear to predate their current symptoms. Cognitive performance was comparable for MCS subjects and controls. The researchers concluded that psychological factors appear to be of central importance in the etiology of MCS. Staudenmeyer et al. (1993) used double-blind provocation chamber challenges to test the effects of odorous air pollutants on 20 individuals presenting with MCS. They found that MCS subjects responded randomly to the chemical challenges.

Around 20 percent of a normal population of college students are predisposed to report symptoms that could be interpreted as those of MCS and that they attribute to exposures to environmental agents (Bell et al., 1993). This suggests the existence of a sizable subpopulation with a propensity to report symptoms that under appropriate environmental and/or behavioral conditions may be manifested as MCS complaints.

The etiology of MCS is hotly debated. The diagnosis of an environmentally induced MCS is generally not recognized by the medical community (Barrett, 1989, 1993; Brautbar, Vojdani, and Campbell, 1992; Orne and Benedetti, 1994; Terr, 1989). Scientific studies of MCS invariably have found alternative psychiatric explanations for the alleged chemical sensitivities. The physiological and biochemical pathways proposed by clinical ecologists are not recognized as valid pathways by traditional medicine. Research evidence strongly suggests that psychological factors play a significant role in MCS cases.

HOW TO TELL SICK BUILDING SYNDROME FROM MASS PSYCHOGENIC ILLNESS

As we have seen, there is considerable similarity between the symptoms of sick building syndrome and those of mass psychogenic illness. As Guidotti et al.

(1987) have shown, it is possible to differentiate between these two types of syndromes. This differentiation is important in choosing what, if any, environmental measures and interventions need to be taken to resolve the situation and in determining an appropriate management response strategy.

The following criteria can be used to distinguish between IEQ and MPI cases:

1. *Is this a sudden incident after a period of few complaints?* MPI incidents tend to occur very quickly, whereas IEQ problems appear to be relatively persistent over time. Burge et al. (1993) surveyed occupants of several office buildings twice within about two years; they found the correlation for the average number of symptoms per worker was a remarkably high 0.96.

2. *Did anything change in the building immediately before the incident, and was this related to the operation of the ventilation system?* MPI incidents frequently seem to be triggered by reports of sudden and unusual odors, whereas widespread odor complaints are seldom reported in IEQ incidents. Struewing and Gray (1990) reported a case of MPI among 1,800 male naval recruits who were evacuated from their barracks because of a suspected toxic gas leak. Upon evacuation, at least 1,000 recruits reported at least one new symptom (cough, chest pain, etc.); 375 had to be evacuated by ambulance from the site, and 8 were hospitalized. But no evidence of any toxic exposure was found. Investigators concluded that the symptoms were the result of physical and mental stress in combination with rumors of odors and toxic gases.

3. *What was the pattern of symptom propagation in the building?* Plotting on a building floor plan the locations of each affected worker, along with the types of symptoms reported and when they occurred, helps to determine whether the symptoms might have arisen from a localized pollutant source, from a pollutant distributed via the ventilation system, or by social contagion from worker to worker. In epidemiology, this is called a *propagation tree.* It is also useful to plot an epidemic curve, that is, the number of cases plotted against time—for example, the number of cases per day or per week for a certain period. This will show whether the incidence of reporting rose sharply and then fell back to a baseline, which is characteristic of an MPI incident.

4. *Is the incidence of complaints consistent with ventilation flow patterns or the heating, ventilating, and air-conditioning (HVAC) system operating schedule?* If the symptoms are worse on Monday than on Friday, this might indicate exposure to microorganisms, such as those that cause humidifier fever. If symptoms are worse in the morning than in the afternoon, this may indicate a ventilation problem in which contaminants accumulate overnight. Giving workers a diary to record their

daily symptoms can help them collect this type of information. A symptom diary also can be a useful way of collecting information on the pattern of symptoms as they relate to possible antecedent events. For example, a worker who consistently develops a headache after lunch may be reacting to food, such as the sodium nitrites in a hot dog. A worker who reports a headache whenever she uses a computer may be experiencing the consequence of inadequate lighting, screen glare, eyestrain, or poor working posture.

5. *Does any recognizable pattern of symptom reporting fit with any known consequences of toxic exposure or a building-related illness such as Legionnaires' disease?* Exposure to a real toxin produces a reproducible clinical constellation of symptoms, whereas symptoms of MPI and IEQ do not show this type of pattern.

6. *What is the rank order of the symptoms being reported?* Lethargy, headache, congested nose, and eye irritation are among the most prevalent IEQ symptoms, but they are not characteristic of MPI. Difficulty breathing, shortness of breath, dizziness, nausea, and odors are characteristic of MPI but relatively uncommon as IEQ complaints.

7. *Did the symptoms decline after the supposed cause was removed or repaired?* In MPI cases some sufferers may eventually develop an apparent intolerance to being anywhere in the building, regardless of any supposed toxic exposure. In such cases, the symptoms invariably have a psychogenic rather than environmental origin.

Finally, it is worth remembering that IEQ and MPI are not clinical illnesses that stem from toxic exposures. Symptoms of IEQ and MPI have not been documented to constitute a serious or life-threatening health hazard. The successful management of IEQ and MPI depends more on managing occupant behavior and the flow of information than on implementing engineering interventions.

ARE PSYCHOSOCIAL FACTORS THE REAL CULPRITS?

People are the best measuring devices we have to gauge the quality of the physical work environment over time, because it is people who remain healthy or get sick. However, as we have seen, people are imperfect devices for measuring both the actual physical environmental conditions in an office and the true health status of their bodies. People will behave in accordance with their own hypotheses about the environmental conditions in a building and their beliefs about their own somatic state. If, for example, people believe the office ventilation is poor and is causing eye irritation, their attentional focus of behavior will change, and they will monitor the eyes more selectively for confirming sensory information. Workers in an office usually cannot correctly attribute causality to their symptoms, and decisions in attributing causality are affected by many cognitive processes (e.g., memory, selective attention, perception, attitudes, and beliefs). How external conditions are interpreted affects the total stress load on each individual worker, and this will vary from person to person because workers differ in their interpretations of the same environmental and occupational conditions as well as their ability to cope with any stressful demands of their workplace and job. This may explain why, under the same prevailing environmental conditions, some workers complain of IEQ symptoms whereas others do not. The results of a number of studies support the concept of multiple risk factors and total stress load in understanding the etiology of IEQ. (Multiple risk factors were discussed in the preceding chapter as well.)

Very few intervention studies have tested the effects of ventilation rate on IAQ and associated problems, but those conducted to date failed to show overwhelming evidence that further increases in ventilation rate have any beneficial effects on the rates of either IAQ complaints or SBS symptoms. Other approaches that complement conventional dilution ventilation, such as source control and the use of localized air-filtration technologies, may produce more significant effects in improving IAQ and reducing complaints and SBS symptoms at lesser expense than increasing the overall building-ventilation rate. Indeed, the use of a breathing-zone filtration (BZF) system, where air filtration is built into the office furniture, can significantly improve IAQ and worker comfort, health, and productivity (Hedge et al., 1993a). BZF may be useful in situations where some symptoms result from temporary physical irritation by fibers and particulates, but this does not undermine the finding that psychological variables also exert a very powerful influence on the reporting of symptoms.

Research shows that building-related illnesses are real but comparatively rare occurrences, whereas IAQ complaints and reports of SBS are much more widespread in buildings but generally are not associated with emissions from specific products or exposure to specific contaminants. Instead, complaints and symptoms are the outcome of the complex interplay of physical environment and nonenvironmental processes and often are initiated by stressful events. Apart from the irritant effects of office dust (Armstrong et al., 1989; Leinster et al., 1990), most studies of IAQ complaints and SBS failed to find significant levels of indoor air pollution. It is possible that one or more undiscovered pollutants are responsible for some complaints and symptoms presently attributed to unknown sources. However, in all studies that have included nonenvironmental variables, good evidence exists that personal, psychological, and occupational variables also affect reports of IAQ complaints and SBS. To date we have only a rudimentary grasp of the processes that actually affect a person's response to indoor air quality.

REFERENCES

Apter A., Bracker A., Hodgson M., Sidman J., and Leung, W. (1994) Epidemiology of the sick building syndrome. *Journal of Allergy and Clinical Immunology* 94, 277–288.

Armstrong, C.W., Sheretz, P.C., and Llewellyn, G.C. (1989) Sick building syndrome traced to excessive total suspended particulates (TSP). *IAQ '89 The Human Equation: Health and Comfort,* ASHRAE, Atlanta, 3–7.

ASHRAE (1989) Ventilation for Acceptable Indoor Air Quality, ASHRAE Standard 62–1989, Atlanta.

Barrett, S. (1989) Unproven "allergies": An epidemic of nonsense. *Nutrition Today,* March/April, 6.

Barrett, S. (1993) Unproven "allergies": An epidemic of nonsense. Report, *American Council on Science and Health,* June, New York.

Bell, I.R., Schwartz, G.E., Peterson, J.M., Amend, D., and Stini, W.A. (1993) Possible time-dependent sensitization to xenobiotics: Self-reported illness from chemical odors, foods and opiate drugs in an older adult population. *Archives of Environmental Health* 48, 315.

Black, D.W., Rathe, A., and Goldstein, R.B. (1990) Environmental illness: A controlled study of 26 subjects with '20th Century Disease'. *Journal of the American Medical Association* 264, 3166–3170.

Bluyssen, P.M., Fernandes, E.D., Groes, L., Clausen, G., Fanger, P.O., Valbjørn, O., Bernhard, C.A., and Roulet, C.A. (1996) European indoor air quality audit project on 56 office buildings. *Indoor Air* 6(4), 221–238.

Bolla-Wilson, K., Wilson, R.J., and Bleecker, M.L. (1988) Conditioning of physical symptoms after neurotoxic exposure. *Journal of Occupational Medicine* 30, 684–686.

Boxer, P.A. (1985) Occupational mass psychogenic illness: History, prevention, and management. *Journal of Occupational Medicine* 27, 867–872.

Boxer, P.A. (1990) Indoor air quality: A psychosocial perspective. *Journal of Occupational Medicine* 32(5), 425–428.

Brautbar, N., Vojdani, A., and Campbell, A.W. (1992) Multiple chemical sensitivities—fact or myth. Guest Editorial, *Toxicology Industrial Health* 8, v–xiii.

Burge, P.S., Hedge, A., Wilson, S., Harris-Bass, J., and Robertson, A.S. (1987) Sick building syndrome: A study of 4373 office workers. *Annals of Occupational Hygiene* 31, 493–504.

Burge, P.S., Robertson, A., and Hedge, A. (1993) The development of a questionnaire suitable for the surveillance of office buildings to assess the building symptom index, a measure of the sick building syndrome. In *Indoor Air '93: Proceedings of the 6th International Conference on Indoor Air Quality and Climate, Helsinki, Finland, July 4–8.* Vol. 1, 731–736.

Burge, P.S., Robertson, A., and Hedge, A. (1990) Validation of self-administered questionnaire in the diagnosis of sick building syndrome. In *Indoor Air 90: Proceedings of the 5th International Conference on Indoor Air Quality and Climate, Toronto Canada* July 29–August 3. Vol. 1, 575–580.

Cain, W.S., and Cometto-Muñiz, J.E. (1993) Irritation and odor symptoms of indoor air pollution. In *Indoor Air '93: Proceedings of the 6th International Conference on Indoor Air Quality and Climate, Helsinki, Finland.* Vol. 1, 21–31.

Chandrakumar, M., Evans, J., and Arulanantham, P. (1994) An investigation into sick building syndrome among local authority employees. *Annals of Occupational Hygiene* 38, 789–799.

Colligan, M.J., and Murphy, L.R. (1982) A review of mass psychogenic illness in work settings. In M.J. Colligan, J.W. Pennebaker, and L.R. Murphy, eds., *Mass Psychogenic Illness: A Social Psychological Analysis.* New Jersey: Erlbaum, pp. 33–55.

Cowart, B.J. (1989) Relationships between taste and smell across the adult life span. *Annals of the New York Academy of Sciences* 561, 39–55.

Dantzer, R. (1993) *The Psychosomatic Delusion.* New York: Free Press.

Doty, R.L., Shaman, P., and Dann, M. (1984) Development of the University of Pennsylvania smell identification test: A standardized microencapsulated test of olfactory function. *Physiology and Behavior* 32, 489–502.

Doty, R.L., Applebaum, S., Zusho, H., and Settle, R.G. (1985) Sex differences in odor identification ability: A cross-cultural analysis. *Neuropsychology* 23, 667–672.

European Concerted Action. (1992) *Guidelines for Ventilation Requirements in Buildings.* (COST 613) Report No. 11 Luxembourg: Commission of the European Communities.

Fanger, P.O. (1988) Introduction of the olf and decipol units to quantify air pollution perceived by humans indoors and outdoors. *Energy and Buildings* 12, 1–6.

Fanger, P.O. (1989) The new comfort equation for indoor air quality. *ASHRAE Journal,* October, 33–38.

Finnegan, M.J., Pickering, C.A., and Burge, P.S. (1984) The sick building syndrome: Prevalence studies. *British Medical Journal* 289, 1573–1575.

Göthe, C.J., Ancker, K., Bjurström, R., Holm, S., and Langworth, S. (1987) Relative humidity, temperature and subjective perception of "dry air." In B. Siefert, H. Esdon, M. Fischer, H. Ruden, and J. Wegner, eds., *Indoor Air '87: Proceedings of the 4th International Conference on Indoor Air Quality and Climate, Berlin (West): Institute for Water, Soil and Air Hygiene.* Vol. 3, 443–447.

Guidotti, T.L., Alexander, R.W., and Fedoruk, M.J. (1987). Epidemiologic features that may distinguish between building-associated illness outbreaks due to chemical exposure or psychogenic origin. *Journal of Occupational Medicine* 29, 148–150.

Gyntelberg, F., Suadicani, P., Nielsen, J.W., Skov, P., Valbjørn, O., Nielsen, P.A., Schneider, T., Jorgensen, O., Wolkoff, P., Wilkins, C.K., Gravesen, S., and Norn, S. (1994) Dust and the sick building syndrome. *Indoor Air* 4(4), 223–238.

Hall, S.K. (1989) The worker with chemical hypersensitivity syndrome. *Pollution Engineering* 76, 79.

Harrison, J., Pickering, A.C., Faregher, E.B., and Austwick, P.K.C. (1990) An investigation of the relationship between microbial and particulate indoor air pollution and the sick building syndrome. *Indoor Air '90,* Vol. 1, 149–154.

Hedge, A., Sterling, E.M., Collett, C.W., and B. Mueller. (1987) Indoor air quality as a psychological stressor. In B. Siefert, H. Esdon, M. Fischer, H. Ruden, and J. Wegner, eds., *Indoor Air '87: Proceedings of the 4th International Conference on Indoor Air Quality and Climate, Berlin (West), August 17–21.* Vol. 2, 552–556.

Hedge, A. (1988) Job stress, job satisfaction, and work-related illness in offices. In *Proceedigns of the 32nd Annual Meeting of the Human Factors Society,* Vol. 2, Santa Monica, CA, pp. 777–779.

Hedge, A., Burge, P.S., Wilson, A.S., and Harris-Bass, J. (1989) Work-related illness in office workers: A proposed model of the sick building syndrome. *Environment International 15,* 143–158.

Hedge, A., Erickson, W., and Rubin, G. (1992) Effects of personal and occupational factors on sick building syndrome reports in air conditioned offices. In J.C. Quick, L.R. Murphy, and J.J. Hurrell Jr., eds., *Work and Well-being: Assessments and Interventions for Occupational Mental Health.* Washington, D.C.: American Psychological Association, 286–298.

Hedge, A., Erickson, W., and Rubin, G. (1993a) Psychosocial correlates of sick building syndrome. In *Indoor Air 93: Proceedings of the 6th International Conference on Indoor Air Quality and Climate, Helsinki, Finland, July 4–8.* Vol. 1, 345–350.

Hedge, A., Mitchell, G.E., McCarthy, J., & Ludwig, J. (1993b) Effects of a furniture-integrated breathing-zone filtration system on indoor air quality, sick building syndrome, productivity, and absenteeism. In *Indoor Air '93: Ventilation, Proceedings of the 6th International Conference on Indoor Air Quality and Climate, Helsinki, Finland, July 4–8.* Vol. 5, 383–388.

Hedge, A., Erickson, W., and Rubin, G. (1995) Psychosocial correlates of sick building syndrome. In *Indoor Air 5,* 10–21.

Hedge, A., Erickson, W., and Rubin, G. (1996) Predicting sick building syndrome at the individual and aggregate levels, *Environment International* 22(1), 3–19.

Hedge, A., and Erickson, W.A. (1996) Associations between cacosmia and SBS in offices, final report. Cornell University Dept. of Design and Environmental Analysis, Ithaca, N.Y.

Hileman, B. (1991) Multiple chemical sensitivity. *Chemical and Engineering News*, July 22, 26.

Hudnell, K., Otto, D., and House, D. (1993) Time course of odor and irritation effects in humans exposed to a mixture of 22 volatile organic compounds. In *Indoor Air 93, Proceedings of the 6th International Conference on Indoor Air Quality and Climate, Helsinki, Finland, July 4–8.* Vol. 1, 567–572.

Iwashita, G. (1992) Assessment of indoor air quality based on human olfactory sensation. Unpublished Ph.D. thesis, Waseda University.

Jaakkola, J.J.K., Reinikainen, L., Neinonen, O.P., Majanen, A., and Seppänen, O. (1991) Indoor air quality requirements for healthy office buildings: Recommendations based on an epidemiologic study. *Environment International,* 17, 371–378.

Knasko, S.C. (1992) Ambient odor's effect on creativity, mood, and perceived health. *Chemical Senses* 17, 27–35.

Knasko, S.C., Gilbert, A.N., and Sabini, J. (1990) Emotional state, physical well-being, and performance in the presence of feigned ambient odor. *Journal of Applied Social Psychology* 20, 1345–1357.

Lehrner, J.P. (1993) Gender differences in long-term odor recognition memory: Verbal versus sensory influences and the consistency of label use. *Chemical Senses* 18, 17–26.

Leinster, P., Raw, G., Thomson, N., Leaman, A., and Whitehead, C. (1990) A modular longitudinal approach to the investigation of sick building syndrome. In *Indoor Air '90: Proceedings of the 5th International Conference on Indoor Air Quality and Climate, Toronto, Canada.* Vol. 1, 287–292.

Meggs, W.J. (1993) Neurogenic inflammation and sensitivity to environmental chemicals. *Environmental Health Perspectives* 101, 234.

Mendell, M.J. (1990) Elevated symptom prevalence in air-conditioned office buildings: A reanalysis of epidemic studies from the United Kingdom. In *Indoor Air '90: Proceedings of the 5th International Conference on Indoor Air Quality and Climate, Toronto, Canada.* Vol. 1, 623–628.

Mendell, M.J. (1993) Non-specific symptoms in office workers: A review and summary of the epidemiologic literature. *Indoor Air 3*, 227–236.

Menzies, R., Tamblyn, R., Farant, J.P., Hanley, J., Nunes, F., and Tamblyn, R. (1993) The effect of varying levels of outdoor-air supply on the symptoms of sick building syndrome. *The New England Journal of Medicine* 12, 821–827.

Menzies, R., Pasztor, J., Leduc, J., and Munes, F. (1994) The "sick building"—a misleading term that should be abandoned. In *Proceedings of the IAO '94: Engineering Indoor Environments.* Atlanta: ASHRAE, 37–48.

Miller, C.S., and Ashoford, N.A. (1993) The hypersusceptible individual. In *Indoor Air '93: Proceedings of the 6th International Conference on Indoor Air Quality and Climate, Helsinki, Finland, July 4–8.* Vol. 1, 549–554.

Morris, L., and Hawkins, L. (1987) The role of stress in the sick building syndrome. In B. Siefert, H. Esdon, M. Fischer, H. Ruden, and J. Wegner, eds., *Indoor Air '87: Proceedings of the 4th International Conference on Indoor Air Quality and Climate, Berlin (West): Institute for Water, Soil and Air Hygiene.* Vol. 2, 566–571.

Nagda, N.L., Koontz, M.O., and Albrecht, R.J. (1991) Effect of ventilation rate in a healthy building. In *Proceedings of the IAO '91: Healthy Buildings.* Atlanta: ASHRAE, 101–107.

Nörback, D., Edling, C., Wieslander, G., and Ramadhan, S. (1993) Exposure to volatile organic compounds (VOC) in the general Swedish population and its relation to perceived air quality and the sick building syndrome. In *Indoor Air 93: Proceedings of the 6th International Conference on Indoor Air Quality and Climate, Helsinki, Finland, July 4–8.* Vol. 1, 573–578.

Olkinuora, M. (1984) Psychogenic epidemics and work. *Scandinavian Journal of Work Environment and Health 10,* 501–504.

Orne, T., and Benedetti, P. (1994) *Multiple Chemical Sensitivity,* New York: American Council on Science and Health.

Otto, D.A., et al. (1990) *Neurotoxic effects of controlled exposure to a complex mixture of volatile organic compounds.* U.S. EPA Final Report 2520.

Otto, D., Hudnem, H.K., House, D., and Prah, J. (1993) Neurobehavioral and subjective reactions of young men and women to a complex mixture of volatile organic compounds. In *Indoor Air '93: Proceedings of the 6th International Conference on Indoor Air Quality and Climate, Helsinki, Finland, July 4–8.* Vol. 1, 59–64.

Pennebaker, J.W. (1982) *The Psychology of Physical Symptoms.* New York: Springer-Verlag.

Pennebaker, J.W., and Skelton, J.A. (1981) Selective monitoring of bodily sensations. *Journal of Personality and Social Psychology 41,* 213–223.

Raw, G.J., Roys, M.S., and Whitehead, C. (1993) Sick building syndrome: Cleanliness is next to healthiness. *Indoor Air* 3(4), 237–245.

Rea, W.J. (1992) *Chemical Sensitivity: Principles and Mechanisms,* Vols. 1 and 2; Boca Raton, Fla.: Lewis.

Simon, G.E., Katon, W.J., and Sparks, P.J. (1990) Allergic to life: psychological factors in environmental illness. *American Journal of Psychiatry* 147, 901.

Simon, G.E., Daniell, W., Stockbridge, H., Claypoole, K., and Rosenstock, L. (1993) Immunologic, psychological, and neuropsychological factors in multiple chemical sensitivity. *Annals of Internal Medicine,* 19(2), 97–103.

Skov, P., Valbjørn, O., and Danish Indoor Study Group. (1987) The sick building syndrome in the office environment: The Danish Town Hall Study. *Environmental International* 13, 339–349.

Skov, P., Valbjørn, O., Pederson, B.V., and Danish Indoor Study Group. (1989) Influence of personal characteristics, job related factors and psychosocial factors on the sick building syndrome. *Scandinavian Journal of Work, Environment and Health* 15, 286–296.

Staudenmeyer, H., Selner, J.C., and Buhr, M.P. (1993) Double-blind provocation challenges in 20 patients presenting with "multiple chemical sensitivity." *Regulatory Toxicology and Pharmacology* 18, 44–53, 1993.

Stenberg, B., and Wall, S. (1995) Why do women report sick building syndrome symptoms more often than men? *Social Science and Medicine* 40, 491–502.

Sterling, E., and Sterling, T. (1983) The impact of different ventilation levels and fluorescent lighting types on building illness: An experimental study. *Canadian Journal of Public Health 74*, 385–392.

Ström, G., Palmgren, U., and West, J. (1993) Microbial volatiles (M-VOC): A causative agent to sick building problems. In E.M. Sterling, C.J. Biera, and C.W. Collett, Eds. (in press) *Building Design and Occupant Well-being in Temperate Climates*. Atlanta: ASHRAE.

Struewing, J.P., and Gray, C.C. (1990) An epidemic of respiratory complaints exacerbated by mass psychogenic illness in a military recruit population. *American Journal of Epidemiology* 132, 1120–1129.

Sundell, J., Andersson B., Andersson K., and Lindvall, T. (1993) Volatile organic compounds in ventilating air in buildings at different sampling points in the buildings and their relationship with the prevalence of occupant symptoms. *Indoor Air* 3, 82–93.

Sundell, J. (1994) On the association between building ventilation characteristics, some indoor environmental exposures, some allergic manifestations, and subjective symptom reports. *Indoor Air* (Supplement No. 2).

Tamblyn, R.M., Menzies, R.I., Nunes, F., Leduc, J., Pasztor, J., and Tamblyn, R.T. (1993) Big air quality complainers—are their office environments different from workers with no complaints? In *Indoor Air '93: Proceedings of the 6th International Conference on Indoor Air Quality and Climate, Helsinki, Finland, July 4–8*. Vol. 1, 133–140.

Terr, A. (1986) Environmental illness, a clinical review of 50 cases. *Archives of Internal Medicine* 146, 145.

Terr, A. (1989) Clinical ecology. *Annals of Internal Medicine* 111, 168.

World Health Organization (1983) *Indoor air pollutants: Exposure and health effects*. EURO Reports and Studies 78, World Health Organization.

Wysocki, C.J., and Gilbert, A.N. (1989) National Geographic smell survey: Effects of age are heterogenous, *Annals of the New York Academy of Sciences* 561, 12–28.

Ziem, G.E. (1992) Multiple chemical sensitivity: Treatment and followup with avoidance and control of chemical exposures. *Toxicology and Industrial Health* 8, 73.

Zweers, T., Preller, L., Brunekreef, B., and Boleij, J.S.M. (1990) Relationships between health and indoor climate complaints and building, workplace, job, and personal characteristics. In *Indoor Air '90: Proceedings of the 5th International Conference on Indoor Air Quality and Climate, Toronto, Canada.* Vol. 1, 495–500.

Zweers, T., Preller, L., Brunekreef, B., and Boleij, J.S.M. (1992) Health and indoor climate complaints of 7,043 office workers in 61 buildings in the Netherlands. *Indoor Air* 2, 127–136.

Chapter 7

Communicating Results of the Investigation

ALAN HEDGE, PH.D.

INTRODUCTION

This chapter builds upon the technical content of the preceding chapters to address the explanation and response phases of communications to those affected by building conditions. All indoor environmental quality (IEQ) problems, sometimes called sick building syndrome (SBS) cases, involve some form of investigation of the building. At one extreme this may be a relatively minor event, involving nothing more than taking a few measurements of indoor conditions in select locations. At the other extreme, it may involve an extensive survey of indoor climate conditions, a thorough analysis of the ventilation system's operation and performance, and a comprehensive questionnaire survey of occupants.

Whatever the scale of the investigation, the need to implement basic principles of good communication remains the same. This chapter presents these principles, along with some examples of dos and don'ts for dealing with the news media. The reader is encouraged to plan, study, examine scenarios, and update existing plans.

ORGANIZING AND MANAGING INFORMATION FLOW

In successful organizations, workers feel involved in the sharing of information with management. The essence of many team-oriented and quality-management practices is the sharing of appropriate information. In any large building, most workers have no idea about the technical nature of building services such as the ventilation system, and most don't really care about these details. What occupants *do* want is to feel safe and secure, assured that the building's environmental conditions are being carefully policed by knowledgeable and professional staff; they want to know that should an environmental problem arise, these same staff members will respond swiftly and effectively. Regrettably, this is seldom the case, even in the best-managed buildings. In the absence of any real understanding of what may be happening in mechanical rooms and within ventilation systems in a building, an employee's imagination can run wild. Any organization owes to its workers and any building manager owes to the building tenants the assurance that basic information on the operation of the building is available to all occupants. This internal sharing of information can take several forms:

• *Newsletter.* Where possible it is useful for an organization to distribute a newsletter that informs workers about the nature and operation of the building's environmental services (heating, ventilation, air conditioning, lighting, etc.). This may be as brief as a single sheet, or it may be a section of a larger publication that is sent to all workers. The newsletter should inform workers in straightforward, nontechnical language about the building's systems, how they work, what environmental monitoring is done, the results of any measurements, any changes being planned, and what to do if environmental conditions are problematic (i.e., who to contact and what to expect). The newsletter should be distributed at least in paper form, although if workers are connected to e-mail, an intranet, or the internet, then an electronic version can be provided.

• *Website.* Workers in many organizations have access to either an internal intranet service or an external internet provider. A company website offers an excellent vehicle for communicating up-to-date information to workers. Such a website should provide workers the opportunity to communicate with building management via e-mail. Analysis of this communication traffic is useful to gauge the opinions of building occupants.

• *Environment committee.* Committees are the organizational lifeblood of most companies. Organizing a committee made up of representatives from management, building-service personnel, occupants, external contractors such as industrial hygiene consultants, and any other relevant personnel (e.g., medical staff) is a familiar and useful way of sharing information and discussing issues. If such a committee meets regularly, then as environmental issues arise, these can be discussed and, ideally, resolved without incident. An environment committee should be set up as soon as an organization occupies a building. Many companies make the mistake of forming such a committee only after an environmental incident has occurred, and in such situations committee members often harbor suspicions and mistrust. The proactive planning of a representative environment committee maximizes the chances that it will be effective in monitoring occupant concerns and dealing with them before any problem becomes widespread in a building.

- *Building tours.* Most building occupants don't know anything about the workings of a ventilation system, and it is easy for them to imagine all kinds of unseen horrors lurking within. Building maintenance personnel are unseen and unknown by most workers. Misconceptions about the potential dangers inside a ventilation system can occur in even the best-run and most-sophisticated buildings. One excellent way to dispel fears and myths about the state of any building is to organize walk-through building tours that include a visit to the mechanical rooms. This gives interested occupants an opportunity to meet with maintenance personnel, to learn about how the mechanical systems operate, and to see firsthand how well the rooms and the inside of the air-handling units are maintained and just how proud and concerned the maintenance personnel are that they are doing a good job to keep the building running smoothly. Once occupants have seen a clean, well-run ventilation system with their own eyes, they are more likely to spread reassurance to others than to stimulate concerns.

- *Complaint procedures.* Most organizations have some kind of formal procedure that occupants need to follow when submitting complaints. Often the occupant simply tells his or her supervisor or phones a building-services engineer. However, occupants often believe that their complaints go unnoticed, and eventually they feel some degree of resentment toward the building-services department; as a result, occupants may abandon the complaints procedure altogether. From the building-services standpoint, complaints about problems with environmental conditions, such as the air being too warm, usually are taken seriously and are responded to. Unfortunately, occupants are seldom told about the response. To avoid disturbing normal work and creating unnecessary questions about the performance of the building systems, maintenance personnel frequently work after hours when the building is empty. In many modern buildings with computer-controlled systems, changes in system performance can be made from a central control room, again without disturbing occupants. Consequently, although actions usually are taken in response to occupant complaints, occupants seldom know about these, and thus they may begin to believe that nothing ever happens when they complain. Giving occupants feedback on the responses to their complaints is an essential strategy in creating and maintaining positive feelings about the building. A simple procedure that we have seen bring about very successful results is as follows: After an occupant makes a complaint about the environmental conditions, the facilities-management or building-services personnel send that occupant a brief note summarizing the nature of the occupant's complaint; what, in nontechnical terms, the staff is planning to do about this; and, if necessary, when and where a maintenance person will be deployed. When the maintenance person has checked and/or adjusted some aspect of the ventilation system, the occupant receives a follow-up note stating what was done and asking the occupant to provide feedback after a specified time (say, one week) as to whether the problem was resolved. This kind of feedback can be provided in writing, by electronic mail, by phone, or in person. This kind of system yields useful building-management information because it can be used to track the performance of building systems and the effectiveness of any interventions. In a building with more than 2,000 workers and a high complaint rate, we found that implementation of this simple feedback system resulted in an almost complete elimination of complaints within 3 months.

When workers suspect that they are getting ill because of something in their building, management's natural reaction is to deny the existence of any problems and to withhold information. These are the worst possible actions, and they usually serve to fan the flames of suspicion rather than douse them. In any SBS situation, good communication practices are essential to effective management and swift resolution. Effective communication describes a two-way process whereby management has an accurate picture of the concerns and attitudes of occupants while sharing with occupants information about conditions in the building in a way that inspires confidence, not anxiety. The methods outlined previously can all be implemented to help to resolve indoor health incidents and, it is hoped, to prevent them. When management shares very sound information with workers, a climate of trust is generated and major incidents are much less likely, even when genuine environmental problems do occur.

RESPONDING TO NEWS MEDIA

Attention-grabbing headlines such as "Death in the afternoon" and "Can your building make you sick?" raise occupant anxiety levels. They also wreak havoc with management confidence, creating understandable caution and mistrust of the media. How the news media report an incident can determine whether a building is evacuated or not. Bad publicity can cost your company substantial amounts of money in lost productivity, absenteeism, and litigation. The mere suspicion that you might have a "sick" building can be enough to bring the newshounds sniffing at your door. Preventing bad publicity depends on how skillfully you handle the news media and how well you plan what and how you will share information.

Information is power; because of this, in many indoor air quality (IAQ) cases organizations actively restrict the information they share with their workers and with the news media, erroneously believing that no news is good news. When a story about your building begins to break, simply saying "no comment" is often the worst thing you can do. Stories about sick buildings have become a fashionable topic in national newspapers and on radio and TV. The term sick building is catchy and evocative. Reporters and the public alike are intrigued by the mystery of most indoor health situations. The fact that suspected hazards in buildings are silent and unseen, combined with the potential audience of millions of Americans who work in offices and could be affected by such hazards, adds to the emotive appeal of this topic.

Reporters and journalists usually are not trained in scientific analysis or building engineering and design. Although the media may strive to present balanced coverage of the situation, often the boundaries between fact and fallacy regarding IEQ become blurred. In the interest of getting a good story that's understandable by the average person, complexities usually get simplified, lay opinions appear as expert facts, and suspicions become conclusions. The lack of definitive answers usually does not detract from the thrill of the mystery, and it seems that the more outlandish the claims made about a building's hazards, the greater the media interest. Sensational media coverage unquestionably can exacerbate concerns among building occupants and fan the flames of panic, sometimes resulting in the premature evacuation of a building for no good reason.

The prospect of dealing with the media can be a daunting experience for an unprepared facilities manager or building engineer. This chapter summarizes sound practices that will help prepare that individual for talking with the media and will help ensure that reports of the building situation are covered accurately and impartially.

The prospect of facing a camera, doing a radio interview, or even speaking with a newspaper journalist can be nerve-wracking. Professionals who work extensively with the media understand the hesitation and reluctance that most people feel. This is especially true where the professional is in a situation where the task is primarily to defend and diffuse concerns about problems in a building. MediaCom, a professional media-training organization, offers the following advice to anyone communicating with the media:

- Always phrase statements simply, clearly, and directly. Where possible, prepare your statements in advance and make a written statement or press release available to the reporter.
- Where possible, ask the reporter in advance what questions you will be asked, and let him or her know what kinds of questions you are unable to answer.
- Ask the reporter questions about his or her knowledge and background and the theme of the story.
- Try to anticipate questions and prepare responses in advance.
- Know the key points that you want to stress, and try to organize these as no more than three or four main bullet points.
- Prepare a theme that accurately reflects your position.

- Never give comments off the record.
- Never opt for the easy "no comment" answer.
- Don't answer questions beyond your knowledge and scope of expertise.
- Don't offer unfounded opinions.
- Don't be irresponsible or flippant in your answers to questions.
- Remain polite, considerate, calm, and pleasant, no matter how much a reporter tries to goad and annoy you.
- Smile and look confident and relaxed in face-to-face encounters.
- Listen carefully to the question you are being asked, and take time to give your answer.
- Make sure that your answers are brief, to the point, and crafted to get the most out of any editing into a headline, video clip, or sound bite.

These key points hold true for successful communication in many situations. Remember that reporters are people who like to be treated as you like to be. Your responsibility in dealing with the news media is to be as honest, straightforward, and accommodating as you can. Also, you need to plan how you can best work with the media to diffuse concerns about your building. In this way you should be able to handle any news situation so that your side is presented fairly and perhaps even favorably. To help you achieve this, the following sections provide you with advice on how to manage different news media situations.

NEWSPAPER AND MAGAZINE INTERVIEWS

There are several types of newspaper journalists, and each works in a different way. Let's say you are currently dealing with a "problem" building. You get a call from a newspaper journalist who is working on this as a news story, possibly even as a lead article for the paper. For such a current news story, the journalist will want to develop an angle that makes the story unusual or in some way sensational. Be cautious and thoughtful about what you say in such a situation; your offhand remarks could be tomorrow's headline. For example, imagine the following exchange:

Reporter: "What about the mystery bug in your building?"
   You: "A mystery bug in our building . . . we haven't found anything yet, but we are still looking, although I don't think we'll find anything."

Your statement could be edited into a headline that reads, "Building manager still looking for mystery bug!", which readers probably will interpret as confirmation that a biological contamination problem does indeed exist in the building. Because this prospect of being misquoted or quoted out of context scares most facilities managers, many opt for a simple "no comment," but avoid this response whenever possible. Sensational headlines are less likely to be used if the journalist is writing a regular feature such as a science column, an editorial, or a magazine piece.

So what can be done when a journalist calls? If you receive a call from a journalist, first politely find out as much about him or her as you can. At a minimum you need to know the following:

- The journalist's name and phone number
- The publication he is writing for
- Whether he is a freelance journalist
- What type of story he is writing—current news, feature article, editorial, and so on
- What she knows about the topic
- What she knows about your building
- What information about your building she has seen—internal reports, memos, letters, and the like
- Who else he has spoken with
- What his deadline is
- How long the article will be
- When the story will appear
- What types of questions he wants to ask
- About how long she expects the interview to last

When you have obtained this information—and it really takes only a few minutes to get it—then we recommend that you tell the journalist that the present time isn't convenient. Arrange a time when the journalist can contact you (allowing sufficient time to meet his or her deadline), even if it's only 30 minutes later; this will buy you some preparation time. If it's more convenient, ask the journalist when you can call him or her, and then be sure to make the call punctually. If possible, arrange to meet the reporter face to face so that you can share graphs, tables, reports, and the like that support your position. Buying some preparation time allows you to focus your mind, to think through your answers to the types of questions that you will be asked, and to address your answers to the theme of the article and the target audience. Also, it gives you time to think about the three or four main points that you want to clearly establish.

When you do speak to the journalist in an interview, start by establishing a time limit—for example, say that the interview must be finished in 10 minutes because you have other commitments. In responding to the journalist's questions, always follow these guidelines:

- Don't interrupt the journalist mid-question, because you might be answering the wrong question in so doing.
- Don't give the journalist additional questions.

- Don't rush your answers; pause after the question has been asked to collect your thoughts before you answer.
- Keep your answers short and specific; don't wander and meander by talking about irrelevant or incidental events.
- Make sure that the journalist understands your answer by using metaphors, anecdotes, and other verbal illustrations to accompany your factual answer.
- Avoid making absolute statements, such as "We've never had a problem in this building," because the chances are that at some time, somewhere, there has been a problem; you just may not have known about it. Be honest and use suitable qualifiers such as, "As far as I know, there hasn't been a history of widespread problems in this building."
- Don't lie or distort the truth.
- Don't make any "off-the-record" comments.
- Don't speculate or agree with any hypothetical scenario suggested by the reporter.
- If you're not sure about facts and figures, say so and request time to check these, or just don't discuss specific details.
- Don't joke about serious issues.
- Remember that anything you say can appear in print.

If the journalist has a good lead time for the article, you may want to refer her to other people that you know and trust to present honest and accurate information about the building. You may also want to offer to mail or fax the journalist some relevant information, such as copies of reports or papers, illustrations, or photographs. A reporter may be willing to send you a proof copy of his or her article so you can check its accuracy, so ask if this is possible. Our experience has been that journalists are reasonable people if you treat them fairly and honestly and help them do their job in a timely manner by providing them with relevant information.

RADIO INTERVIEWS

If you get a request for a radio interview, follow the preceding guidelines—find out as much as you can about the interviewer, the story, the show, and the radio station. Radio interviews can take various forms: edited sound bites, recorded interviews, live radio-phone interviews, and studio interviews. Edited sound-bite interviews usually are limited to 60 to 90 seconds, and these are then distributed for use by stations across the United States. Edited radio news interviews can also be short (1–3 minutes), though if it's a radio show you may be aired for 5 minutes. Live radio-phone interviews involve you in answering questions by phone; your answers are aired directly and unedited. Studio interviews involve you going into a radio studio. The studio interview might be recorded for subsequent editing or aired live. In preparing for a radio interview you should generally follow the principles already outlined, although the uniqueness of radio requires that you follow some unique principles for the best results:

- Where possible, keep your answers short and to the point, no more than 10 to 20 seconds long.
- Don't rush your answers, and avoid using "fumble words," such as "um, that's a good question, now . . .," or inane phrases such as "at this point in time . . ." (just say "now" or "at present").
- Refer to the interviewer by name when you respond.
- Don't speak too quickly. A person who's nervous tends to speak faster; this makes listening more difficult.
- Don't speak in a monotone; try to sound lively, energetic, and confident. Use changes in the quality of your voice to emphasize points and feelings. For example, speak more slowly and with a lower tone to denote the seriousness of a situation; speak louder (never shout) and more forcefully to impart a sense of urgency or support a point of view that you want to convey. If the interviewer asks you a question in a frenzied voice, reply in a calm and relaxed voice to imply confidence and stability in your handling of a situation. It is a good idea to practice these techniques using a tape recorder before the interview takes place.
- Don't cite too many facts and figures or present data to irrelevant levels of accuracy (e.g., "fifty-three-point-two-four percent of those interviewed said . . ."), because it is difficult for listeners to remember these and follow the theme of your answer.

Chances are that you will be sitting down when you are interviewed, so make sure that you sit upright so that you will be the most alert and will have the greatest vocal range.

If you are conducting a live radio-phone interview, try to make sure that you are in a setting that is as quiet as possible. Do the following:

- Close your office door.
- Tell your secretary to hold all calls, or turn your phone off.
- Turn off any audible signals from devices like pagers, digital watches, and the like.
- Turn off any fans, room air conditioners, printers, or computers.
- Don't try to drink coffee, tea, or soda or eat during the interview; you may get caught with your mouth full.

- Don't try to type on your computer while you're talking.
- Hold the phone so that you talk across the mouthpiece and not directly into it. This helps to eliminate hissing and popping sounds.

If your interview is being taped for subsequent editing and you fumble or give a clumsy verbal answer to a question, ask if you can record over your answer. When the interview is finished, ask if you will be able to preview the edited version before broadcast, by either listening to it over the phone or reviewing a written transcript.

TELEVISION INTERVIEWS

Television is the most powerful medium because it combines words, images, and action. Most people in the United States, and even the world, get their news from the TV. Because of this, competition for airtime is intense and stories typically are short. While you may have spent hours with a reporter, the news story that airs will have been extensively edited, and typically it will last only between 30 and 90 seconds. Although brief, such an interview combined with suitable images can be either damning or exonerating.

Key points in preparing for a TV interview are the same whether your interview is prerecorded or live. However, in a prerecorded situation, try to get to know the reporter and develop a rapport with him or her. Ask if you can stay involved by checking over any final scripts or even previewing the video clip. Be as helpful as possible to the reporter, and never be overtly critical of his or her work. Also, make sure that you look the reporter in the eyes and that your body language is in agreement with your verbal answers. Above all, keep your answers brief and to the point; never forget that whatever you say and do will be edited, so try to ensure that even with severe editing your message will not be lost. To illustrate this, consider the following example:

> *Interviewer:* "You've just evacuated your building. Do you really think that there are no problems with your building?"
>
> *Answers to avoid:* "No, what I'm saying is not that there isn't a problem but that the problem isn't a serious one." (This may get edited to a "No" answer, which implies that you agree that there is a problem.)
>
> "That's a loaded question, but I'll do my best to answer it . . ." (This may get edited to "That's a loaded question," and you will appear hostile and evasive.)

"No comment." This answer is guaranteed to raise suspicions that you're evading the truth and that you're untrustworthy.

Possible answers: "Yes, I think that the building is fine. We've evacuated people just as a precaution while we thoroughly check everything again."

"Yes, my experts have given the all clear for us to reoccupy the building. We evacuated only as a precaution until we had the full details of the situation. We always act in the best interests of our workers." (Even if these answers get edited to simply a "yes," you've made your point that you think the building is safe).

"Ambush" Television Interviews

Watch any TV news program and you'll see example after example of an unsuspecting interviewee confronted by a reporter with a microphone and camera. A favorite technique of investigative reporters, the ambush interview is designed to catch you off guard, when you are most likely to panic and say something spontaneously. In such interviews the reporter might ask outrageous questions in an attempt to provoke an emotional reaction that can be viewed by millions. You can, however, prepare yourself for ambush interviews in several ways:

- If your building becomes a hot topic, think about where a reporter is most likely to try to conduct an ambush interview. If your building has good security, then most likely this will happen as you arrive at or leave the building, or as you arrive at or leave your home. The interview might happen just as you get into or out of a car. If you say "no comment," most likely your response will be aired and your reputation will be questioned.
- When you've thought about where an ambush might take place, mentally prepare yourself for it. For example, as you leave the office, start to think about what points you would like to make if you are interviewed. If you rehearse your three or four "must-air" points, you will be able to effectively deliver these to a reporter whenever and wherever you are approached.
- Remember, the ambush interview is no different from a regular TV interview, but the reporter tries to use the element of surprise to catch you off guard.

Above all, if you are ambushed, don't panic; remain calm, smile and appear amiable, and maintain good eye contact. Also, in this situation you can take the initiative to end the interview whenever you want. Don't just walk away—tell the reporter that you must leave, for a good reason; perhaps even thank them for the interview; smile; turn; and then confidently walk away.

FORMAL NEWS CONFERENCES

In many situations, you can structure media interest in your viewpoint by organizing a news conference to which you invite reporters. In a typical news conference, you can start by stating your point of view; you can supplement this by providing reporters with supporting information such as prepared news releases and copies of documents, reports, graphs, tables, photographs, and so on. During the first part of any news conference, you can control the information flow. If you simply want to present your viewpoint to the media, you might choose to end the conference then and there, without taking questions. If this is your chosen strategy, have someone else open the conference and introduce you, and then close the conference after you have finished speaking. This deflects questions from you to an administrative person who should not be expected to have the knowledge required to answer them.

If you want to accommodate reporters, you can elect to answer their questions after your statement. In this case it is a good idea to ask reporters to submit their questions in writing in advance, so that you can have all the necessary facts and figures on hand to answer them. Consider giving reporters a written copy of prepared answers to their questions. It is also a good idea to set a time limit on how long you will answer questions. You might also want to limit the number of questions that any reporter can ask, as well as the scope of questions. Don't get drawn into answering spontaneous questions shouted out from the floor, unless you are comfortable doing this.

If you choose to take unscreened questions, you might want to use a "buy time" tactic so you can assemble your thoughts into a good spontaneous response. Such tactics include asking the reporter to repeat the question, asking for clarification of the question, or rhetorically repeating the question back to the reporter, perhaps with a request for clarification. When you are being asked a question, look directly at the reporter, but once you begin your answer, look around at the entire group of reporters. Don't tolerate interruptions to your answers. If you are interrupted, pause, remain calm, and either pick up where you left off or begin answering the question again. If you are asked personal or hostile questions, don't react to them, but reiterate your points of view. As you approach the end of the question period, have your administrative person tell the reporters that there is only enough time for another two or three questions. At the end of the questions, remember to repeat your three or four "must air" points. In the news conference you are the person who manages the flow of information, so take the best advantage of this privileged position.

Preventing Panic

All the advice in this chapter should help you manage information flow in a way that prevents occupants from panicking about potential problems in a building and prevents the news media from fanning the flames of panic. Otherwise, a building-incident situation can quickly escalate, and your building can be evacuated before any evidence of threat has been collected and analyzed.

Should such a situation arise in your facility, the planning and preparation you have undertaken can help reduce the severity of the incident's impact. With regard to building problems, you should develop a plan that specifies who will be with responsible for certain actions and that coordinates management information between different parties involved in any incident, such as human resources, the facilities department, building-service engineers, external liaisons, medical assessment and treatment personnel, and the litigation staff. Because events can occur very rapidly once an incident has begun, maintaining good communication between all parties can become extremely difficult. Your plan should detail at least the following:

- Who is responsible for assembling and processing specific information about events in the building
- Who should undertake specific testing in the building
- Who decides what action to take to implement any engineering controls
- How and when medical services should be involved
- What information will be communicated to occupants, how it will be communicated, and who will do it. This strategy should include efforts to minimize rumors and the spread of any unnecessary concerns.
- What information will be communicated to the news media, how it will be done, and who will do it
- What information will be recorded about this incident and who will record it

Unfortunately, few building managers publicize near-miss incidents in which their buildings were almost evacuated, and in many cases where evacuation has occurred litigation is ongoing, so it is not easy to find publicly available documents that tell you how others have coped in similar circumstances. With a plan in place, however, you will be better prepared for any future incident, even though it may never arise.

Chapter 8

Government Inspections of the Building

James T. O'Reilly

Practical Issues

At some point a government agency will become involved with workers' complaints about the building in question. The very pragmatic advice that has evolved from this author's years of interaction with government inspectors is simple: though private companies *do* have legal rights to resist government inspections (as discussed in the following section), it is more prudent to allow the inspection while making sure that the inspector gets it right and does not make a major sampling error or erroneous written observation.

The building owner's in-house or contracted safety professional probably is far better equipped to deal with indoor health allegations than the government representative who responds to a telephone complaint from one of the workers in the building. A certified industrial hygienist or other safety professional is more likely to obtain accurate measurements of levels of the materials cited in the complaint than is the local health department inspector whose primary area of experience may be cleaning practices in restaurant kitchens. Even the federal or state inspector who responds to a complaint is likely to have difficulty with an indoor air allegation because of the variable factors that must be examined in the short time available before the next inspection assignment. Helping this person get it right the first time is a very wise use of the owner's resources, since any comments in documents generated in error by a government inspector can be, and often are, referred to as the gospel truth if and when the building complaints grow into litigation. The ring of truth attaches not to the content but to the source.

The right way to deal with an inspection is to augment the government inspector's technical capability with that of a consultant or in-house person with equal or better credentials. The safety professional will urge the public employee to hold off on reaching a conclusion in the report until a set of supporting measurements can be assembled with the proper equipment. Sharing of measurement machines or properly calibrated equipment by the two or more investigators can make the results more credible and accurate, even if the results are not entirely favorable to the preferred claim that no problem exists. By cooperatively measuring and performing the tests that are appropriate for the type of illness that is being alleged, the owner is following all the right steps to make the conclusions as factual as possible.

LEGAL RIGHTS

The Fourth Amendment to the U.S. Constitution forbids government agencies from making "unreasonable searches" on private property. A series of Supreme Court cases concluded that the right to be free of inspections on a commercial property is limited but not eliminated. The commercial-property owner has the legal right to refuse consent to an inspection. The government official then must ask a court, such as a local magistrate, to issue a warrant that allows the search to go forward. Once the warrant is issued, the owner must allow the search to proceed or risk being jailed for contempt of court. The property owner's lawyer, however, can ask the magistrate or judge to halt the inspection as illegal or unreasonable, or the lawyer can later try to suppress the use of the evidence gathered in an enforcement-type inspection.

The facility search warrant is issued if the government has either a complaint of a health concern that requires investigation, which is the usual situation in sick building syndrome (SBS) cases, or an "administratively neutral search plan," a program that looks at classes of buildings rather than selecting an arbitrary site. When a complaint is involved, the supporting affidavit that the inspector files usually claims that a health problem exists and that the inspection is needed to protect the workers' health. In those cases where the company resists and tries to have the warrant denied, government-informant testimony alleging company misconduct is likely to cause significant adverse press coverage.

Because most inspections are performed with the owner's consent, these constitutional privacy rights and limitations issues usually don't get to court. Only if the site has been involved in controversy that generates adverse government reactions and the owner suspects retribution or bias should the owner prepare for litigation over the conduct of an inspection.

How the Inspection Is Conducted

Government inspectors want pieces of evidence that they can easily incorporate into their view of the facts. The less helpful and more secretive a company is, the less likely it is to find receptivity for its presentation. But that does *not* mean that every piece of documentation needs to be delivered to the regulatory agency.

Reasonable methods of inspection should be used. An SBS or indoor environmental quality (IEQ) claim should not be investigated by unreasonably intrusive methods. The taking of samples, placement of air monitors, and so on, should be done with minimal disruption. The inspector usually arrives without prior notice. Because of the health issues involved the government inspector's arrival can be viewed by employees and the media as a solution to a problem; the owner or manager should therefore be careful in how an objection is made (typically with the help of the owner's attorney). The owner's objections to inspection could degenerate into a real credibility challenge. It is extremely important that the person who hosts the inspector's visit use good common sense.

Occupational Safety and Health Administration (OSHA) investigators will routinely insist that they be allowed to interview workers without management representatives present. This arrangement intrudes into the management relationship, but it's probably not worth the legal effort to object, since workers can talk to OSHA by phone or from their homes, and the owner wants to avoid the appearance that the company has something to hide.

The National Institutes of Occupational Safety and Health (NIOSH) also conducts inspections of workplaces; its purpose is to add to the base of information from which its scientists can produce criteria documents regarding particular health risks. NIOSH evaluations of work sites are research-oriented rather than punitive. The NIOSH inspection tends to be part of a larger effort to assess workplace risks.

Required Reports

The government inspector will anticipate that the required log of injuries has been maintained at the workplace. OSHA regulations compel the company to maintain a log of "reportable" injuries, which include fatal and severe injuries, worker injuries, resulting in lost work time, and other serious harms. The point of debate between worker advocates, such as unions, and building owners or managers will be whether the health complaints of headaches and sensitivities qualify as OSHA-reportable.

Chapter 9

Indoor Health Litigation
and the Building Management

James T. O'Reilly

INTRODUCTION TO LITIGATION OPTIONS

This chapter summarizes the legal options for the building manager or safety professional who confronts a health or safety allegation in the form of a legal claim for damages to an individual. Any indoor health claim is serious; the owner and the owner's representatives should treat the allegation as the first step in what may become a complex legal proceeding. Ideally, the claim never gets to the lawsuit stage, because the technical and medical interactions described in earlier chapters have been successful.

If the claim results in a lawsuit brought by one or more of the building's tenant companies or their employees who are building occupants, a series of questions are likely to arise that should be understood as early as possible in the conflict. Of course, not all claims ripen into lawsuits, and not all lawsuits are complex; but the better-prepared manager is aware of the possibilities and prepares as early as possible for the eventualities that might happen.

This chapter presents the series of questions that often are asked about health-allegation lawsuits. Your own attorney may well have a different view; these responses are based on general experience and study of the field rather than being tailored to any specific situation.

WHY WOULD ANYONE SUE?

Lawsuits allege a type of injury that justifies the court's award of some damages. Typically, lawsuits over indoor health conditions claim a physical injury to the respiratory system or other physical health conditions of the individual worker. Those suing usually assert that they incurred medical expenses, lost wages from sick time when they felt unable to work, experienced pain and suffering from the health condition, and in some cases lost the value of their services by being forced to change to a different locality or type of employment.

Some building-related lawsuits brought by tenant companies allege a breach of contract for rental of the space, because the space is alleged to be unusable by virtue of the environmental health condition. The landlord's failure to provide a "habitable" work space that is free from the offending condition is alleged to be the basis for damages, including the costs of moving, the rent paid while the space was untenable, and the higher costs of replacement space.

Other lawsuits, brought by building owners, assert that the design or construction of the building breached contracts or breached implied warranties, a form of contractual obligation that arises from relationships formed in the planning, engineering, and construction of a building. Architects, heating and ventilation engineers, and construction contractors are the typical third-party defendants when the building owner tries to pass along some of its potential liability.

A final set of defendants provide furniture, carpets, lighting, and other products. It can be alleged that use of these products caused adverse health effects, for example, a new rug emitted odors into a confined space or a new space heater vented fumes without adequate exhaust mechanisms or warnings. State laws governing product liability usually differentiate between these tangible products and the fixed asset, the building in which the products are placed. State laws that limit liability for product-related injuries might or might not also cover the building itself.

What Does the Indoor Health Claimant Need to Prove?

The person who sues a building owner or other defendant for indoor health–related problems usually needs to establish that the unsafe or defective condition of the building was a *proximate cause* of a documented injury incurred by the individual plaintiff. The elements of a claim of negligent operation would be that the person operating the building had a duty to protect occupants, that the owner knew or should have known of the risk, that the risk was the proximate cause of the claimant's injuries, and that a reasonable building owner would not have acted as the present owner did.

For example, assume that fumes in a 12-story building were alleged to have caused asthma for 30 workers on two floors. The claimants would have to show that the defendant owner knew about the presence of the fumes, and they would have to present medical evidence of the asthma; these would be shown by medical records, air monitoring and sampling reports, and other records. Experts would be called upon to show that causation was more than a mere remote coincidence and that the fumes proximately caused the harm. Testimony would be offered showing that more-reasonable owners of similar buildings had acted to avoid similar types of injuries.

WHAT STEPS SHOULD BE TAKEN WHEN A LAWSUIT IS THREATENED?

Like it or not, the defense of a lawsuit alleging indoor health risks is a group project. The individual or company that owns building, along with advisors, is a principal participant but does not make all the decisions. The owner's insurance carrier is very interested if coverage for such claims is allowed and not covered by the standard pollution exclusion. Lawyers for the company will explore the coverage of applicable insurance contracts, and if a continuing exposure to unhealthy air has been alleged, then the prior years' insurance policies may need to be examined to determine whether several insurors may be jointly called on for the defense.

The first step for the prudent building owner is to work with lawyers to investigate the facts that underlie the potential lawsuit. The preceding chapter described this process of fact-gathering. A competent industrial hygienist with the right measurement equipment should be assigned to the task, preferably one with experience in responding to indoor air complaints. If the complaints are similar in separate locations, sampling each of the locations and one site without complaints is recommended, so that a comparison and base line can be obtained.

The first-line supervisors and managers who know these particular complainants should be consulted very promptly, because they may be able to provide valuable background facts explaining the sudden appearance of the allegations, or they might also have noticed a deficiency in the work area's ventilation or other on-site conditions.

The next step is to consult with expert advisors on indoor air quality problems, whom you may be able to find through a trade association, professional society of engineers, or state chamber of commerce. The patterns of dispute avoidance that they recommend might accommodate the complaining persons without the high transaction costs of a lawsuit.

If a union represents the workers, management should consider asking the union to call in its own health professional to work with the company's professionals in examining the scene. Many larger unions are knowledgeable about workplace exposure issues; at a minimum, the company gains some credibility if the external review of the potential risk is handled by a worker-paid professional.

Because insurance companies are important funding sources for damage-claim settlements, giving the insurance company prompt notice of the claim is essential. There are benefits to acting early, even if the language of the policy calls for notice once the suit papers have been served. It may be that the insurance carrier's loss-prevention experts can help with specific ways to relieve the problem. Sometimes, coverage is denied immediately on the grounds that indoor contaminants fall within the pollution exclusion in standard form contracts. The insurance company may be aware of some method of assuaging concern and alarm that has worked in other contexts.

In a few unusual circumstances, after trying lesser measures to deal with the facts and allegations, the company may wish to consult with federal or state Occupational Safety and Health Administration (OSHA) personnel to explore all the possibilities. A decision by these agencies that no health risk exists would be a valuable assertion for the defense of a lawsuit. OSHA's regulatory policies, enforcement mind-set, and reputation of arbitrariness regarding workplace health enforcement are currently in transition. It is feasible to expect that some of the expertise OSHA and its sister agencies, the Environmental Protection Agency (EPA) and the National Institute for Occupational Safety and Health (NIOSH), have gained in other indoor air situations would help the building owner. The risk is that the enforcers may penalize the company, and their adverse findings then would strengthen the workers' civil-court action.

What Steps Should Be Taken When the Suit Is Filed?

Notify your lawyer and your insurance carrier by phone, with copies of the suit to follow as soon as possible. Your lawyer will begin researching the allegations with the help of your team. Immediate insurance notification is generally required as a precondition for the insured party's claim for the protections of the insurance policy.

Consider the risk of a plaintiff's misuse of the lawsuit as a basis for generating bad publicity. If the possibility exists that the suit will be publicized in the news media, have a public relations person standing by to offer the defense's perspective if and when questions are received.

Notify the first-line supervisors of any active employees who are named as plaintiffs in the suit that no discipline, discharge, or action that looks like retaliation can be implemented without prior consultation with the defense lawyer.

Preserve any documents that relate to specifications of the building design, its ventilation system, controls, and filters, and so on. If the claim alleges that carpeting caused the problem, find and preserve records of the order and any warranty information that the carpet vendor may have supplied at the time of installation. Many companies routinely discard documents after a certain amount of time; through direct messages to line supervisors and staff experts, management should suspend any further document destruction and should preferably centralize copies of all documents with the defense attorney.

Who Are the Defendants?

Identities of the defendants will vary with the nature of the claims being made. In every case, the building owner can expect to be sued, and wherever it is permitted by exceptions to workers' compensation laws, the actual employer of the worker may be sued as well.

If a product is alleged to have given off fumes that caused respiratory problems, the building owner, the product maker, and the vendor of that product will each be named. State procedures and limitations vary, but it is likely that strict liability for harm from a product will apply only to the actual manufacturer of the challenged product.

If a poor ventilation system is alleged to have existed, then the designers, architects, and installers will be sued instead of a product manufacturer (the makers of filtration or air-exchange controls might be added in a sophisticated lawsuit that addresses the specific weaknesses of a site's ventilation).

When Is a Class Action Used?

Courts recognize that in some circumstances, the same health risk was incurred in the same or very similar work sites by more than one person. If the numbers of potential litigation plaintiffs make it unwieldy to take each case to a jury one after another, then the plaintiff's lawyers will ask the court to allow them to act for a whole class of persons who may have been exposed.

Class actions are favored by the person suing the building owner because the presence of multiple claims has a persuasive, cumulative impact on juries' perceptions about the health risks. A jury that hears of 100 persons claiming to be affected is more likely to award a large verdict; since it might excuse a single plaintiff's claims as merely idiosyncrasies, but may perceive that 100 affected people can't all be wrong.

Defense counsel usually opposes class actions. Humans' individual physical responses to stimuli are so variable that it can be unfair to lump all claims into one pool. Medical testimony about airway problems will of course be influenced by which person had a heavy smoking habit, which person was formerly an asbestos installation worker, and which person had been treated for chronic asthma before being hired at the site.

Some class-action suits allow the plaintiff group to speak for all the affected persons, for example, all workers in the same building, with no opportunity for the persons to file their own separate suits. These mandatory class situations are relatively infrequent. Most allow an *opt out* period, during which a person who is in the class can choose to file his or her own suit for the injury. The person who opts out will not share in the class settlement but may be able to win an individual damage award that is larger than the class members' payments from the defense.

What Is the Document-Discovery Process?

The gathering of evidence against a building owner often begins with its own files. *Discovery* means the phase of the lawsuit in which the files of one side are opened to the review of the other (subject to exclusions, known as *privileges,* that shield certain categories of records from compelled disclosure). In federal practice and in many states, the presumption is that any record that is likely to lead to the discovery of useful evidence can be obtained, even if the document itself would not be of much value in the court case.

The typical document-discovery process involves a detailed file review by the companies and persons who are the target defendants. Files consist of both the company's or institution's official files and the individual files kept at the desks of multiple managers, service providers, and supervisors. Documents are copied and marked with code numbers to make it easier to index the set of records, and in some cases bar codes are used in case a computerized index might be needed.

Decisions about which records are not to be shared are made by the trial defense team. Privileged documents are listed in a log or listing of the types of documents for which claims will be asserted against the plaintiff's request.

Will My Deposition Be Required?

Oral questioning of witnesses before the trial is called the *deposition.* The defense offers its lists of potential witnesses, and the plaintiff brings in its experts and other witnesses. Each side can call the other's potential witnesses for depositions to determine, prior to trial, what they are likely to say in trial testimony. Depositions are intimidating events, but with sound preparation and a skilled attorney advising the witness, they can facilitate the process of settlement rather than causing further problems for the affected managers.

Do Lawsuits Usually Reach a Jury Verdict?

No. A small percentage of lawsuits reach the jury; most are settled after months of negotiation that end with a release of all claims in return for a cash payment. How the turning point is reached, and how the settlement is achieved, varies from case to case. Factors in the settlement include the amount of insurance coverage that is likely to be available, taking into account exclusions and denials of coverage; the size of the plaintiff's money demands; the likely effect of settlement on other lawsuits; the appearance of good or harmful evidence as seen by either of the negotiating sides; and other factors peculiar to the dynamics of the relationship between the owner, the employer, and the individual plaintiff.

Chapter 10

Litigation by the Tenant's Employees

JAMES T. O'REILLY

LIABILITY RISKS FOR THE BUILDING OWNER

This chapter focuses on the liabilities of the building owner to employees of a tenant company who claim they were affected by sick building syndrome (SBS) or a related illness. The defendant in such a case is the owner, not the employer/tenant. This is because in most cases, state workers' compensation barriers to civil lawsuits, discussed in Chapter 11, preclude suits against one's employer.

This section does not address tenant companies' suits against building owners for inadequate maintenance, since these are conventional lease disputes rather than complex SBS negligence actions. Health-related building-environment claims are a breed apart from normal lease problems.

Tactical choices arise immediately: should the owner sue the tenant—that is, should the tenant be named as a codefendant by a separate cross-complaint between the owner and the tenant? More often, cooperation and the long-standing relationship between the tenant to this building is more important, so avoiding such a suit is prudent.

The key issue in the SBS case will be proving that the illness was caused by a specific condition of the building. The tenant's acts often play a major role in the causation of a building-related health allegation. There are two ways to dig into the tenant's role in the factual causation: (1) cooperating to find the causal connection, regardless of who is more at fault (if any "fault" exists); and (2) if the tenant is uncooperative, suing the tenant to drag out the information through pretrial discovery and depositions.

Cooperation makes more sense in the majority of cases, especially where the relationship between the building owner and the tenant company have been positive. SBS is a people problem, and it can be addressed through people-to-people negotiations. The owner does not want to elevate the individual employee's suit into a major controversy. The owner does not want to make the tenant evacuate the building to better defend itself in a suit brought by the owner. Continued use and activity in the building by the tenant also helps the owner, since it undercuts the employee's claim that the building is so dangerous that its airborne contaminants breached the implied warranties of habitability (discussed in a later section).

In practice, cooperation makes a huge difference in the speed and adequacy of the investigation. The owner's consultant professionals who wish to investigate the work space, for example, an industrial hygienist or a team of ventilation and infection specialists, are better able to analyze the potential causes with full cooperation of the tenant/employer. The cooperating tenant/employer may be asked for a list of chemicals used in its photo-developing lab, hiring and assignment histories for the individual plaintiffs and a control group of similarly exposed workers without these alleged symptoms, and any information about recent renovations or specialized cleaning compounds used in the maintenance of the tenant's space. It is difficult enough to perform a competent assessment, given the "needle in the haystack" nature of SBS investigations, without the added burden of resistance and hostility from the management of the tenant company.

Suing the tenant will probably erect a barrier that diminishes the accuracy of the investigation. Lawyers for the defendant tenant/employer will decline voluntary cooperation and begin accusing the owner of full responsibility for SBS cases. The specialist consultant team that serves as the owner's "detectives" needs to search for clues about causation by inquiring, experimenting, testing, and reviewing records of the tenant's space. Interviews with the plaintiff employee's immediate managers are critical diagnostic tools. So it would be futile to try to sue this tenant while the owner needs and expects the tenant's complete cooperation.

Legal Theories

Negligence

Three primary legal theories are available to the employee who sues the building owner: negligent acts that caused physical harm, breach of warranties of habitability of the rented space, and assault by involuntary exposure of the employee to a harmful airborne contaminant.

Negligence of the building owner in maintaining a clean indoor environment is likely to be the major grounds for the lawsuit by the tenant's employees. The four elements to be established are the existence of a duty of care between the owner and the employee of the tenant company, breach of that duty, injury of the employee, and proximate causation of this injury by the act or omission of the building owner. In lay terms, the lawyer will show that the owner was supposed to supply a reasonably safe place to work, the owner failed to do so, the worker became ill, and the illness was caused by something in the building.

The legal theory of civil tort liabilities allows a negligence claim if the challenger can establish each of the four elements of duty, breach, injury, and proximate causation. Existence of a duty to the individual employees by a building owner is not obvious. A duty does exist to maintain the premises so the tenant company can operate safely, for example, the building complies with fire codes, has proper wiring, has exit doors and fire alarms, and so on. The lawyer for the SBS plaintiffs argues far more: that the owner of the building owes a duty to safeguard the sensitive physical conditions of individuals who work in the building.

This claim of a specific duty of the owner to avoid airborne contaminants that would harm the sensitive individual employee would, in most cases, have to be inferred from general principles of law. That is because the lease contract between owner and tenant probably does not identify the workers who will be present or even refer to employee health matters at all. Disclaimers in the contract are likely to work against such a health claim.

The legal question under state law of whether any duty of due care exists to protect the ability of the employee to breathe indoor air without contaminants is likely to be decided in pretrial motions, since it is not a factual dispute for a jury. (Legal duty is a question for the judge; whether or not contaminants existed and caused harm is a factual issue for the jury.) Case law of the state where the suit is pending might make it very difficult for an individual to win, especially where the employer is not suing the owner and other workers are still using the building without interruption.

If the building owner and the tenant/employer each feel the complaint is invalid, perhaps unique to this individual making the complaint, the response to the suit will probably assert that no duty exists beyond that of protecting the reasonably sensitive, reasonably healthy worker/occupant. The fact that the tenant's other employees continue to work in the same space without adverse effects helps the defense to deny that a duty to extremely sensitive persons exists. Part of the defense will focus on medical records of the employee's past experiences with allergies and irritations.

Once a duty is shown to exist, the plaintiff has overcome a formidable legal barrier, but it is not the last one. Next the claimant will need to show that the duty was breached by something that the building owner/defendant did or failed to do. Breach means that the action was contrary to the duty, or the omission was in violation of the duty to perform some act. For example, if the trial judge decides as a matter of law that the owner of the building owed a duty to plaintiff X, an employee of tenant Y, then a breach of this duty would consist of actions like redirecting the chimney of the building so that furnace smoke blew into Y's offices and affected X. Because causation is very difficult to establish and the symptoms claimed in SBS cases tend to be vague, the lawyer for the employee will need to isolate the date(s) on which illness was observed and then focus on which action breached the duty: Was it errors in ventilation design? Were air filters changed too infrequently? Was carpet adhesive installed on September 4?

Correlating the timing of the claimed illness and the timing of the actions of the owner/defendant may be difficult. The gradual appearance of a set of symptoms that are cumulatively damaging to the plaintiff worker does not make for an easy case of breach. The vagueness of the onset of symptoms, first noticed in June 1997 but reported to doctors in March 1998, for example, makes it hard to isolate the breach action that occurred back in May or June 1997. Vague feelings that "We just got sick" and "The building was just too tight that winter" do not satisfy the plaintiff's burden to show that the owner of the building had breached a duty of due care.

The third of the four measures of negligence is the existence of an injury. Chapters 4 and 5, which deal with diagnosis, will help explain the difficulty that a plaintiff in an SBS claim must face in asserting that an injury occurred. It is important to recall that vaguer symptoms with more-tenuous connections to the workplace are not persuasive as identifiable injuries; the claimant must connect an injury to a particular breach.

The fourth and highest barrier of all is the proof of proximate causation. This means that it is more than 50 percent likely that there was a direct causal connection between the act or omission that was a breach of duty, and the specific injury. A lawnmower made with a factory error that left its safety switch disconnected causes a foot to be cut off; proximate causation by the factory error is easy. An office building air-conditioning duct that is damp might or might not have proximately caused the flu symptoms of a fifth-floor clerk; the plaintiff presents physical evidence and expert testimony about causation, and then the jury decides whether there is sufficient proof of a close, proximate connection.

Breach of Implied Warranty

The second theory for potential employee litigation, breach of the lease's implied warranty, is bound up in the contractual duties of owner and tenant under the lease. It is not clear that an employee of the lessee can sue the lessor for damages. Under state contract law of warranties, the lessee's employee would be asserting a right to health protection under the lease, but the injured person is not a party to the lease, and that document probably does not mention health or safety issues. If the lease excludes all implied warranties by means of a disclaimer, or contains terms that allocate responsibility to the tenant for the maintenance of its space, then the challenger is likely to be unable to win a warranty claim. If the space is still being used, but the person who is suing does not feel able to work there, the warranty of habitability is not breached, since the individual is not named in the lease as a person subject to protection under the lease.

Warranty claims also involve the complexity of timing. The injured person has a duty to give notice of breach of warranty soon after it is discovered. The worker who suffers an adverse effect but waits to sue may have missed the deadline by which a notice of warranty breach must be brought to the owner's attention. Delay may bar the breach-of-warranty lawsuit entirely.

Assault

The third of these claims is that the owner, without consent, subjected the individual to physical assault with fumes or dust. This is very unlikely to succeed at trial, since juries have a colloquial view of assault that does not include simple maintenance lapses in a building's mechanical systems. This theory is sometimes used for egregious hazardous chemical spills onto another person's property. A typical assault claim in an SBS case would allege negligent omission of cleaning that left contaminants in the air ducts or building environment. This is not the intentional and direct adverse impact against a person that the term assault usually connotes.

Class Actions

The most cumbersome and expensive form of damages litigation is the class action, an effort to combine multiple persons' claims against a single defendant. This is done in a way that achieves easier court handling of common aspects of a product-safety or building-safety concern. Class-action lawsuits could be brought, for example, by five employees of building tenants who allege that they represent all other employees in the building.

Using a class action saves time and effort for the court, but only if it avoids what will truly be repetitive claims. If the claims are not repetitive because they represent different types or causes of injury, then a class action should not be used. The class of employees needs to have very similar claims of physical injuries and common claims of causation before their lawsuits can be consolidated into a class.

Building owners' lawyers generally prefer that no class action be brought; individual employee claims should be handled individually, they reason, because the costs of each case will act as a barrier to potential plaintiffs. The danger for owners from class actions is that it cost-justifies damage claims that individually would be too small to serve as the basis for lawsuits. Many people would choose not to pursue a trivial complaint through the court system if they had to individually convince a jury that they were harmed. The great majority of complaints relating to SBS tend to be of transitory, minor physical illness. Instead of fighting 10 serious court cases with 10 sets of proofs and defenses, building owners could face a class-action request for 800 individual workers, some of whom had illness, but the majority of whom experienced little or no inconvenience.

Plaintiffs' lawyers prefer to convince a court to grant these small claims the legal status of class actions, and then to settle all of the claims for a single amount. By parceling out the shares of the pool of damages so that everyone involved gets some amount, the class action gives the most seriously ill persons less money and gives somewhat of a windfall to persons who ordinarily would either not have sued or would not have won a sizable jury verdict. The legal fees that come out of this pot of settlement funds tend to reward lawyers so richly that the class-action approach has come under considerable negative scrutiny. Class actions can be criticized as coercive blackmail perpetrated by the legal system against the targeted companies.

How Should the Building Owner Interact with Other Defendants?

The building owner may be a joint defendant with the companies that sold carpet, installed wallboard, painted, or equipped the building with ventilation machinery. This approach will team the building owner with one or more codefendants for as many types of exposures as the plaintiffs can allege. The real target is the building owner. The cost of challenging a service provider or equipment vendor is less than the cost of challenging a building owner, since the owner will resist and the vendor will probably pay something to avoid the costs of litigation. The vendor's share of damages may be so small that a skilled plaintiff lawyer who seeks a settlement can sue a dozen separate suppliers or service providers who had some contact with the indoor air conditions. Then the lawyer can use the cash nuisance settlements received from each to fund the primary litigation against the building owner.

The building owner in an SBS case should be careful to document the vendor's awareness of the claim and its participation in the meetings at which joint defenses are discussed. Records are very important, especially if the plaintiff's lawyer is seeking a series of settlements. Lawyers who are familiar with such meetings warn of high levels of mistrust and suspicion among the codefendant companies. A vendor that made a $500 profit or a service contractor that cleared $1,000 on an installation deal will not be willing to pay $20,000 in legal costs to stay active in the defense of SBS allegations for a building that will not be likely to continue a sales or service relationship. In short, the costs of the lawsuit are very likely to drive vendors and contractors toward a prompt settlement, even if their conduct would not have been objectively wrong or their products deemed objectively dangerous to users.

To reduce the anxiety and costs of defending the SBS case, the building owner (and the insurance carrier, if coverage of SBS claims has been accepted) should map out a legally privileged document, offering a joint defense agreement for the several interested parties on the defense side. Typical joint agreements allocate the eventual jury verdict award, if any, in percentage terms that are more attractive to the smaller defendants in return for their pledge to continue in defense of the suit without separately paying a settlement to the plaintiff. The chances are that the liabilities will be borne by the building owner and perhaps one or two of the vendors of equipment or services. Other parties to an agreement get the benefit of some assurance of a cap on their outlays (if the case is lost at trial) and are not exposed to the huge expense of being left standing alone as the last defendant facing a possible punitive jury verdict.

The building-product vendor's warranties and instruction materials will be very important to the plaintiff's case based on implied warranties. It may be that a major equipment purchase negotiated between a sophisticated seller and building owner deviated from normal boilerplate contract exclusions and disclaimers. A lawyer for the prudent building owner who expects an SBS claim to ripen into litigation should examine the promises made by the ventilation-equipment maker, paint supplier, or carpet manufacturer as early as possible in the creation of a defense file.

THREATS OF ADVERSE PUBLICITY

When SBS complaints come from employees of the tenants and ripen into a well-publicized lawsuit about illness from environmental conditions within the building, the savvy building owner can anticipate that a strong public-relations response will be needed. Television coverage of sick workers evacuated from the premises will destroy the the building's marketability among realtors and potential local renters. This publicizing of claims of negative health effects harms the site's economic viability.

To avoid this, a building owner's press statements should be coordinated with those of the tenant company to strongly disagree with the assertions. A delicate balance needs to be drawn, as Chapter 7 on press communication tactics, explains. At this point it is usually too early to give a definitive causation statement; saying, "We don't know why X and Y are ill" provides fodder for news media representatives focusing on one specific person's complaint. A government official's expression of belief that the building is safe is much more desirable but may be harder to achieve.

In the building owner's press conferences about the litigation defense, the media can be reminded that very few of the SBS allegations made in headlines have later proven to withstand scientific scrutiny. As a category—even though this particular case is still undecided—SBS claims are so easy to make and so hard to prove that caution is appropriate for the reporters who cover this particular story. As Chapter 8 suggests, genuine sensitivity, concern for the individual, and desire to get the right medical assessment should be expressed, because these portray a serious attentiveness to workers' concerns. The press spokesperson for the owner should, however, directly address the myth that every SBS charge is true. Allegations of SBS are people problems and will be dealt with sensitively.

The most delicate of public relations challenges are publicity stunts asserting SBS as part of an unrelated campaign. A rival company's promotion of its own building leases by attacking the safety of a competitor's office building could be an extreme case of trade libel if the allegations are false.

But, for instance, a union dispute that distorts facts to assert SBS as a basis for organizing workers against the tenant/employer produces a harsh side effect. Union charges can work against the rights of the building owner. The owner may not be able to make headway in the media against the union allegations; the rebuttal rights allowed an employer during a union campaign are stringently limited by federal labor laws. If the allegations suddenly disappear because the union wins its desired contract, then no correction is issued, the building owner does not get back its good reputation, and press attention is sparse for the factual story that finally appears stating the building is *not* causing real health problems for workers. Vindication is cold comfort, since many who read the front page today will not see the column buried on page 17 three months from now that explains that the building was judged safe. The building owner who expects a challenger to use scare tactics should immediately engage a defense-oriented public-relations firm to manage its response.

Chapter 11

Legal Options of the Complaining Occupant Employees

JAMES T. O'REILLY

INTRODUCTION

This chapter explains how the employees who work in the building may act on their complaints of building-related illness through the legal system. The particular area of most activity is likely to be the workers' compensation system. Lawsuits alleging negligence by the owner and other problems were addressed in Chapter 10.

In future years, when more enforcement activity occurs under the Americans with Disabilities Act (ADA), the ADA will also be a significant factor. As awareness of the full effects of ADA litigation spreads among lawyers who represent complaining employees, more interaction will need to take place between the two sets of defense lawyers—those who work on issues of disability that are traditionally labor-related, and those who deal with the defense of civil tort liabilities for property owners.

HOW WORKERS' COMPENSATION INTERACTS WITH INDOOR HEALTH

The workers' compensation (WC) system was designed to remove litigation conflict, uncertainty, and transaction costs from the handling of workplace accidental physical injuries such as slips, falls, severed limbs, and back strains. Workers would not need to sue, and employers could hold down the transaction costs of on-the-job injuries through a predictable insurance system managed or closely regulated by the state. Accidents involving workers would produce a predictable result within a well-understood system.

The claim for WC benefits based on indoor air exposures typically asserts that the worker had a prolonged exposure that caused the injury, and the exposure was a greater hazard than that faced by the general public. In cases asserting occupational disease, the worker needs to show actual causation of the disease by conditions that are characteristic of that occupation, that the disease was actually contracted during employment in the particular occupation, that the actual conditions of employment were characteristic of and peculiar to a particular occupation, and that the occupation presents a risk of an unusual illness or a higher risk of a commonplace illness than the general population experiences.

The employer's personnel department often engages a WC service provider within an insurance company to provide the WC processing needed for claims. The typical claim is for a nontrivial on-the-job burn, strain, or bone injury that results in a lost-time injury of several weeks' duration. The company's safety professional, who may be the company's own risk-management specialist or the insurance company's contact person, works with the WC contact person. The WC contact person manages the medical follow-up, any filings required by the entity that oversees WC programs in that state, and any follow-up paperwork that is needed. The safety professional should work with the WC manager to investigate how the claim occurred, for example, when a machine guard was not secured, and whether preventive action should be taken. Good cooperation between the safety professional and the WC manager is essential to hold down the company's costs of lost-time accidents.

IMPORTANT ELEMENTS OF WORKERS' COMPENSATION

Several elements of the WC system are important for the building owner/manager to understand.

1. WC is an *exclusive remedy*. This means it is a trade-off form of compensation—workers get assurance of financial compensation and medical care for the on-the-job injury but give up their right to sue the employer. The good news is that fewer lawsuits result. The bad news for the nonemployer building owner is that the secondary players in the situation, who are not the employer, *can* be sued by the injured person. So an office-building tenant's employee who claims physical harm from sick building syndrome (SBS) may be barred from suing the tenant but is not barred from suing the building's owner.

2. WC pays primarily for accidents and not for intentional injuries or long-standing workplace conditions. With some exceptions, WC administrators and courts require that the worker show a specific time, place, and cause that produced the accidental injury. One of the exceptions to this norm occurs when the employee complains of an occupational illness that is more prevalent in the industry than in the general population. No accident needs to have taken place; if there is a cause-and-effect relationship to this type of illness within this type of work, and the worker shows the illness arose from workplace exposure, then WC systems are more likely to pay. But the particular causal connection does not yet exist in indoor health complaints to achieve WC payments. And an illness that arises often in the general population, such as smoking-related lung cancer, is less likely to be compensated.

3. Workers' compensation is a lenient, proworker method of compensation that is never as fact-driven or cautious as the civil jury system's traditional process of basing decisions on adversarial proof: Presumptions of compensability arise once the worker shows that there could have been a causal connection between the workplace and the illness or injury. WC hearing officers tend to grant the worker's claim of work-related injury if the assertion has any legitimate basis. The employer then has the burden of eliminating any reasonable possibility that workplace conditions were a factor in causing the injury, or presenting an alternative cause that would exclude the workplace exposure as a substantial cause of the harm.

The significance of this principle is that the building owner who adamantly defends the safety of the building's working conditions, and who would defend it in court whenever challenged, might be unpleasantly surprised when a state hearing officer or WC screening official grants compensation based on a presumption or on less-than-perfect proof of causation. The owner should fight the claim after becoming armed with the technical facts. The better line of reasoning is that office-type workplace environments do not cause compensable injuries because exposures to smoking, paint, carpet, common dust, and so on are routine parts of everyday life, independent of exposure during the person's working hours. It is not enough that an exposure was caused by the place of employment; the disease must result from the nature of the employment. If the initial decision goes against the owner, it is *not* a slap at the building's safety in the same way that a court decision might be deemed a verdict of guilt. The lighter standard of proof in WC cases makes it more likely that borderline claimants win. The decision should be appealed after the full administrative record of owner evidence is entered into the agency's hearings record.

4. Workers' compensation provides built-in incentive for instigating lawsuits against owners other than the individual's employer. Proving workplace causes of breathing difficulties or other sensitivities is difficult. Those who win, especially those employed in more highly paid occupations, are often surprised

at the small amount of money they receive from WC after the payment of legal fees. The injured person may begin to seek other targets from whom money is available. Once the worker hires a lawyer and the search for a deeper-pocket source of additional income begins, it is not likely that the building owner's potential liability will be overlooked.

5. WC payment decisions have a group effect on other workers that cannot be taken lightly. The employer should contest adverse decisions, because silence stimulates the filing of other claims. Precedents that are implicit or explicit can be set when an employer decides not to contest the award of WC payments for a claim of work-related injury. Implicitly, other workers will believe that the grant of compensation represents an admission or official conclusion that the company and/or building owner indeed harmed the worker. That is *not* what a WC award means, but workers without much awareness of the system may assume so. Explicitly, state insurance-fund managers and hearing officers will tend to follow as precedent a decision allowing injury claims by one employee, so that others alleging similar injuries are also granted compensation. The clustering of allegations of injury that occur in SBS cases, discussed in Chapter 10, is a phenomenon that makes it essential that the employer contest any injury allegation or sensitivity claims that the employer doubts. Compassion for the individual employee should *not* be manifested by lax handling of injury claims.

6. Remedial action *after* the WC investigation has concluded is important in avoiding subsequent "me-too" claims by other employees. The employer's ability to correct the condition varies with cost and other factors; the employer who has paid once for the accidental injury or occupational illness should then take preventive steps that are visible to other workers to reduce the temptation for other workers to pile on additional claims.

7. Of all WC cases, those involving chemical sensitivity may be the most difficult for claimants to establish. If the employer declines to accept the WC claim, the employer can expect to face litigation that is not precluded by the availability of a WC remedy. Denying the claim opens the door to a possibly long and expensive lawsuit.

CAN EMPLOYEES SUE
DESPITE WORKERS' COMPENSATION BARRIERS?

In general, workers alleging a right to workers' compensation payments for accidental injury due to SBS are not able to also sue their employers for damages. Workers' compensation is generally the exclusive remedy for the employees, once the WC system is shown to apply. The two options for the worker are to sue the building owner, if different from the employer, or to claim the exceptional condition known as an *intentional tort* in a suit against the employer.

Intentional torts are physical injury cases in which the employer injured the worker either directly, such as a small-business owner assaulting his or her employee in a fit of anger, or indirectly, where the employer was more than negligent, had knowledge of a risk, and then consciously and intentionally exposed the worker to that risk, with the specific expectation that the employee was substantially likely to be injured. This latter category has been described as a conscious and deliberate intent directed to the purpose of inflicting an injury. Either circumstance will proceed outside WC because it is outside the class of accidents to which WC is directed.

Intentional tort is the allegedly injured person's vehicle of choice for sick building claims when the employer is also the owner of the building. A tiny percentage of workers' compensation claimants bring these suits, and few of these are successful. The barrier to such lawsuits remains very high, and courts have no incentive to widen the number of such disputes that can be brought into the civil courts.

Proof of the employer's knowledge and intent about the risk is needed in the intentional tort claim. This is assembled from testimony, statements, and measurements of workplace conditions. Elements usually presented to the court are that the employer had actual knowledge that an injury was substantially certain to occur and that the employer willfully disregarded that knowledge. If the injury is substantially certain to occur as a consequence of actions the employer intended, the employer is deemed to have intended the injuries as well. States have debated legislative reforms of their WC systems in recent years, and changes have tended to tighten the intentional tort category—for example, moving knowledge to "actual" knowledge and the test of injury from "substantially certain" to occur to "certain" to occur. In states with the "certain to occur" norm in their laws or court decisions, the SBS litigant would need to show that for its employer, no doubt existed with regard to whether the injury would occur.

One of the starkest examples of an intentional indoor air quality injury case was the criminal prosecution of a film-processing company that hired only people who could not read English-language warnings on chemical containers. The workers complained often about the inadequate ventilation around vats of hydrogen cyanide gas. Chemical labels gave adequate warnings, but the workers could not read them; eventually, one worker died and several others were poisoned by the cyanide.

Defense lawyers representing the building owner who receives an intentional tort claim need to develop a rebuttal argument, assembling proof of care and prudent preventive action to rebut the claim of knowing risks. Where available, showing that the owner made corrective efforts will undercut claims that the injury was known to be substantially likely to occur.

When defending the owner, lawyers should anticipate that the intentional torts defendant will assert that supervisors, first-line managers, or building-maintenance managers caused the problem, for example, by ordering office workers to close windows and work in dusty cubicles. A lawyer can anticipate that the defendant will try to hold the company accountable for supervisor misconduct. The same high threshold applies to suits against the employer or its supervisory personnel; most such lawsuits are against the employer, with its deep pockets, rather than against individual managers.

Proving the condition of the building caused injury is the principal barrier the challenger needs to overcome. Did the condition of the building cause this physical effect? Were the same exposures occurring elsewhere? In a New York case, a worker alleged unsuccessfully in court that exposure to many different chemicals in a hospital had adversely affected her health, making her so allergic that she could not work. Her WC benefit claims were denied because no accident had occurred and the illness did not arise from the nature of her particular job functions. Exposure to allegedly harmful chemicals was held by the court not to be an accident or a disease that would naturally and unavoidably result from an accidental exposure.

After that barrier is overcome, the worker needs to prove a knowing and intentional exposure to a known risk. Did the building possess a risk of injury in its interior conditions that was substantially likely to cause physical harm to this worker, and did the defendant know of that risk? In a Florida case, Anchor-Hocking Corporation was sued by a factory worker who alleged the company diverted a smokestack so that fumes went into a building instead of outside, and that the employer periodically shut off the ventilation system altogether. The courts allowed that intentional tort claim to proceed and held that it was not barred by WC remedies.

The second element of an intentional tort is a command to the worker to enter or remain in a hazardous area, regardless of high risk of harm. What did supervisors know? Did supervisors order these workers to stay in what they knew was a dangerous location? Conflicts in testimony need to be resolved. Giving prompt attention to health-related complaints is a prudent occupational plan for preventing these concerns.

DEALING WITH THE COMPLAINER: LIMITATIONS

Should employers try to solve the sick building allegations by quickly firing the person who complained? Probably not! Such an action exposes the employer to large liabilities, bad publicity, and suspicion of callous disregard for worker safety. So it is virtually never recommended as the method of addressing indoor health problems.

The complainer may be correct, and the building systems may be out of adjustment or otherwise contaminated. An unforeseen condition may have affected air intakes, for example, as other chapters of this book have indicated. The employer should follow several steps:

1. Health-related building allegations are not simply an engineering concern. Personnel problems are best handled by a human resources professional who is sensitive to the legal problems associated with today's regulation of labor-force issues. Managing people is not as easy as it once was, and minefields of legal protections and limitations can be dangerous for the well-intentioned manager. The team approach can include the building manager, the safety professional, and the human resources professional, with the worker-employer interactions handled by the human resources person to avoid possible miscommunications that can worsen employee relations.

2. Documentation is a critical piece of the company's defense. Records of how the employee's situation was handled will make a great deal of difference if a lawsuit ever arises. A human resources specialist will emphasize the need for giving written, documented responses to the individual and recording what complaint was addressed on what date with what measures. Taking notes at meetings and adding them to the file also helps prevent future memory lapses.

3. Some workplaces are subject to laws that protect the complainant against firing or discipline in safety-related matters. If a complaint, regardless of its merit, was made to a designated safety or health agency under the laws that apply, then a discharged worker can claim back pay, reinstatement, and legal fees against the employer. These whistle-blower laws apply to many government offices, and states or localities may have created similar protections for private-sector workers. The purpose of the laws is to encourage safety-related communications without fear of retaliation. Of course, the downside of the laws for employers is that a chronic complainer who has fabricated or exaggerated threats to health is likely to be bolder if the individual knows that the employer cannot respond with discipline. The details of how a worker gets protected as a whistle-blower and how agencies investigate relevant charges are beyond the scope of this book, but your employment lawyer should be aware of them.

The sociology of modern office environments is also beyond the scope of this text, but whistle-blowers who must be kept on the payroll pose a particular challenge. The worker who complains to the Occupational Safety and Health Administration (OSHA) by means of an anonymous phone call or letter is protected against any repercussions. The prudent employer should treat the complaint as a warning that one of its employees either has a serious health concern or is likely to be a source of future difficulties that require internal remedies, so as to induce the person to resolve the concerns without government intervention. The group dynamics within the workforce may change after indoor complaints, since other employees will be watching for the supervisor's and manager's response. The employer will naturally dislike the complaining person's agitation of problems after a factual basis for the complaint has not been found. But the supervisor cannot halt the conduct with discipline, because safety complaints made to OSHA are a federally protected activity for which no discipline can be imposed, even if OSHA finds no violations.

PRETERMINATION CONSIDERATIONS

Termination of the complaining person is possible, but only after the employer has made efforts to deal properly with the complaint. A well-documented file should show that the person was told of the factual variance between the complainer's view and the technically accurate and medically correct facts. Most states have kept the traditional doctrine of "employment at will" for nonunion employees. Termination at will does not require proof of grounds for dismissal, except where one of the exceptions to the doctrine exists. Many of today's senior managers grew up with that informal expectation during their careers. What makes the indoor health case different?

In today's climate of expensive employment litigation, decisions to terminate the problem employee in a sick-building case should be discussed with a competent labor-relations lawyer before the decision is implemented. Pitfalls of discharging the employee without a legal check of the decision can include back pay orders, demands for reinstatement of the hostile ex-employee to the same position held at the time of firing, and sometimes damage awards, with additional attorney fees.

Discharge for reasons that are *against a public purpose* can invalidate the decision to terminate, and result in costly settlements or court orders. This exception means that the worker told others about the health problem for the purpose of avoiding a hazard to their health and by doing so had aided a public goal of health-risk prevention. That public purpose created an exception to the employer's traditional freedom to terminate the person at will.

UNIONS AND THE COMPLAINING WORKER

Unions represent a small minority of private-sector office workers, the traditional sick-building claimants, and a somewhat larger proportion of state and local government office workers. The presence of a union entirely alters the employment at will concept.

Once a union has signed collective-bargaining agreements with the employer, the employer is expected to meet and confer with the union about such matters as health concerns in the workplace, complaints about working conditions, and terminations. The discipline process is either spelled out in the contract, referenced from some preexisting employment manual or program, or (in the case of public employers) spelled out in civil-service rules or their equivalent. The typical union contract constrains the employer's right to discharge members of the union's bargaining unit (workers covered by the contract), and it grants the worker a series of grievance hearings before the termination is deemed final.

Some unions include labor-management *safety committee* provisions in the contract. Where this exists, the employer must allow the union to participate fully in the decisions made about studying indoor health allegations, evaluating the risks and remediating the problems in the building. The meetings of this group add to the delay of decisions and can increase the employer's consulting costs; but such a committee offers a way to enlist the credibility of the union's technical expert regarding the methods by which assessment, evaluation, and remediation are being handled.

In some cases, unions can use indoor health concerns as a threat in their drive to organize office workers; such fear-driven campaigns are unfortunate since the fearful worker may not wish to listen to objective measurements and technical presentations. The building owner or employer who is undergoing a union-organizing campaign should be especially quick to respond to allegations of risks. Being unresponsive in the face of a challenge is merely adding fuel to the fire of the organizing effort. But the converse is not always true; responsive and factual rebuttals will not convince those who have a particular agenda to which the indoor health rumors can add emphasis. If a union-organizing campaign is under way, the employer probably has a competent labor lawyer working on the defense side, and that professional should become very involved in the ways in which building measurements are communicated to the workers.

Even if a union does not represent the affected set of workers, a potential land mine exists in the concerted effort of multiple employees to refuse to work under unsafe conditions. The group decision to collectively leave the workplace and avoid the risk of exposure they fear could qualify as a safety strike, which is protected by federal labor laws, even if the workers do not at the time have a union representing them. Reinstatement and back pay orders may take years to achieve, but the employer's decision to fire the group may turn out to be an expensive choice.

WORKER COMPLAINTS TO GOVERNMENT

The regulatory investigation power of OSHA and its state counterparts in about half the states in the nation is applicable to most private-sector workplaces.

(Public-sector employees typically fall under a state-law equivalent program administered by state officials.) OSHA and its research colleagues in the National Institute for Occupational Safety and Health (NIOSH) can become involved in some indoor health cases but lack the resources to become involved in many of the claims. OSHA sometimes adopts standards and penalizes employers who violate them, but no specific OSHA standard for indoor air was in place as this text went to press.

The complaining person's assertion of a physical injury poses a particular problem in the current legal climate of protection for those who make claims of discrimination under the ADA. The employer must first understand what ADA requires in terms of an accommodation before deciding whether to oppose the claim that a disability exists. If the amount of the individual accommodation is small and the changes will not set a precedent that is difficult to live with, then the owner should consider offering an accommodation. The more costly, precedent-setting, and disruptive the accommodation, the less likely it is that the company will voluntarily make the changes for the complaining person. The lose/lose situation for building owners that results could potentially provoke other employees to make their own separate demands.

The ADA claim of failure to accommodate an employee's physical disability is best handled by the company's employment counsel, who is likely to have experience with claims of disabled persons about hiring, transportation, and other related employment issues. The counsel should not take lightly a demand for accommodation; even though the content of the claim often seems out of line with the average person's expectations, federal entities like the Equal Employment Opportunity Commission and the Department of Justice take the ADA requirements more literally.

THE EMPLOYER'S ROLE IN THE EMPLOYEE'S LIABILITY ACTION AGAINST VENDORS

When one or more employees sue the suppliers whose building materials or maintenance items are alleged to have caused the harm, what role should the employer play?

Of course, if this level of discontent can be avoided, it should be resolved before litigation begins. The lawsuit may distract employees from their work, and the company's productivity may suffer during the deposition, pretrial, trial, and appeal phases of such a private suit. Litigation discovery and development of the injured persons' claims will take days of person-effort with no return to the employer. And a situation that has led to litigation is likely to be a continuing source of problems, since an employee should not be expected to be happily productive in a building within which working conditions are so bad that he or she has found it necessary to sue. As Chapter 6 of this text discusses, the effect of dealing with the anger and malaise of multiple discontented employees is more powerful than that of responding to one employee's flu symptoms or allergic reaction.

A natural tendency exists for employers *not* to want their employees distracted by avoidable litigation. But if the employer actively blocks the lawsuit or discharges the person who sued, to appease the angry building owner, then the result may be that the ex-employee brings in the employer as a codefendant in the lawsuit, with results that are equally negative. A form of mediation or negotiation of solutions is far superior to litigation, and the sympathetic employer can use its economic weight to aid the cause of the aggrieved employee.

The employer should first evaluate the claims of the individual's physician or health advisors. Beginning with the specifications and purchase documents for building materials, construction, or operating systems like heating, the employer should work with the building owner to gather expert laboratory measurements and technical data on potential sources of the problem. If a measurable contaminant is found, the employer and owner should document jointly the amount and source before the building is cleaned and scrubbed, so that before-and-after measurements can be presented in the event of persistent complaints.

Serious consideration must be given to the risk that the employer will be dragged in as a third-party defendant in the case. Say, for example, that workers claim a vendor of building materials was strictly liable for injuries caused by off-gassing of the product. The legal standard for strict liability requires a showing that the product that caused the injury was in the same condition as it was when it left the factory. The best tactic for a defendant may be to claim installation errors or failure to follow the right directions. So the employer's records and witnesses will be deeply involved in the case. If the employer had chosen to sue the vendor itself, then the employee's participation would have been fully justified.

Chapter 12

Complaints by the Tenant Company

JAMES T. O'REILLY

HANDLING COMPLAINTS BEFORE LITIGATION

This chapter discusses claims of building-related illness made against the building owner by the tenant/employer, who alleges that their employees became ill or were harmed by the employees' workplace conditions. The allegation is easy to assert but hard to prove, as discussed in other chapters of this text; the claim is more likely to be credible to the owner when it is brought by a corporate tenant with some sophistication (or who has retained a consultant of some experience) than when the claim is made by an individual employee.

All health-related claims must be taken seriously and handled with empathy. But the complaint from an official of a formidable tenant, who acts rationally and presents a detailed expression of the belief that risks exist in the workplace environment, is a significant event. That kind of communication deserves the highest levels of attention from the owner's management team. Common sense suggests that tenants will receive more attention if they approach the owner with some facts and a basis for the charge; a tenant's rush to accept employees' suspicions without further investigation is not a sign of the cooperative approach that owners welcome.

CLAIMS OF ECONOMIC DAMAGE BASED ON OWNER'S BREACH OF LEASE

The terms and conditions of the lease are a starting point for disputes between tenant and landlord over the safety of a rented building space. Terms guaranteeing such conditions as "quiet enjoyment of the leased premises" are common, but many others can be used as well. These terms govern the legal aspects of the tenant's demand that an owner pay to remediate the conditions, or the tenant's demand for forgiveness of rent during the time that the remediation of the health concerns is under way.

A disclaimer releasing the building owner from certain liabilities may have been included in the contract. Even so, the prudent landlord does not simply assert lease clauses and refuse to help when the tenant appears to have a legitimate concern. Sensitivity to the delicate position of the tenant, who has complaining employees but no control over building operations or design, can alleviate some of the tension inherent in an indoor environmental health case. The economic loss of having tenants evacuate or abandon a building, along with the harm that this can cause to the landlord's reputation, might not be covered by the landlord's insurance policy, so an effort to help the tenant by sharing information makes sound business sense.

While the tenant and the landlord are gathering data and engaging in informal discussions about alternatives, the lawyers working with the building owner will examine the lease's disclaimers, hold-harmless, or indemnity clauses to determine the maximum amount to which the landlord could be exposed in claims by the tenant. At the same time, the owner and its insurance carrier should have preliminary contact to discuss the potential losses. Chapter 14 cautions the reader that insurance carriers can be expected to disclaim coverage if they consider the building contaminants causing the alleged illness to be pollution.

Commercial leases for office space tend to be tightly written in favor of the building owner, but a tenant with sophistication and some clout in negotiation might have added contract language with some nonstandard representations and warranties. If this modification of terms occurred, a possible misrepresentation claim might surface in tenant-versus-landlord disputes, for example, if the tenant's negotiator had won some written assurance about the fitness of the building for a particular use. But this becomes an issue only where that use is relevant to the indoor health concerns. Awareness of disclaimers, exclusions, and the atmosphere created by exchanges of letters should all be factored into this decision about possible litigation.

The owner is not always on the defensive. Tenants are not entirely free of liabilities in indoor health cases. Where a suspected cause of the illness condition in the workplace was a modification that caused damage to the premises, for which the tenant is responsible under terms of the lease, the building owner might deny any liability.

An irrational tenant's sudden evacuation from the building warrants a strong rebuttal. The owner might threaten to sue the tenant for any damages the owner incurs in responding to the indoor health allegations, including loss of value of rental leases that result from tenant-generated bad publicity. Of course, the prudent landlord wants to keep its tenants happy and will not sue a tenant who is still in the premises, except under extraordinary circumstances.

WHEN HABITABILITY WARRANTIES ARE IMPLIED

The implied warranty that premises will be *habitable* is a legal matter of contract law. It is implied from the contract relationship even if the contract is silent. But any implied warranty can be disclaimed by a proper phrasing of the lease agreement. An owner's disclaimer may exclude liability for any lack of habitability for a particular purpose, for example, no promise is made by the owner that its office building is habitable as a location for a chemistry laboratory.

The indoor health allegation by one or a few workers usually does not mean that the site cannot be occupied and used by others who are not claiming sensitivity problems. The terms of the state's contract law for similar property owner–tenant disputes should be considered, but a landlord's indoor health case may be the first one in your state to test whether habitability is violated by one or a few workers' complaints of sensitivity.

A creative lawyer for the tenant can try to expand habitability to 100 percent worker satisfaction. That claim asserts that the building should have been useful for all of the employees who were brought into the building by the tenant/ employer, but it was not useful because some of the employees experienced symptoms and claimed illness. Thus, it would be argued, the lease should be implied to have promised utility for all occupants, yet the building was not totally habitable because of the complaints. The flaw in this reasoning is that it stretches what can be implied as an extra promise in a contract that is silent. When typical courts are asked to imply a missing term into a contract, they look to other case precedents and then ask whether the agreement implied a term that was not otherwise included in the contract. Courts will not be as generous as the plaintiff would like in adding terms not otherwise present. And if the prudent owner has included a disclaimer of implied warranties, this line of argument disappears.

DAMAGE CLAIMS FOR INTERRUPTION OF WORK

What happens to the owner's economic position when the tenants' workers panic and stop working in the building? One of the most expensive elements of a tenant's assertion of losses due to a "sick" building is the claim for loss of income while the workers were distracted, or in some cases evacuated, because of the building's harmful conditions. The owner may have to face a claim, for example, that $200,000 in productive work output was not performed during the days that the employees inside the building were complaining of a strange and harmful odor. If the workers called ambulances and the media and refused to reenter the building, the employer's work was not getting done. The owner will not have an easy time rebutting these quickly mounting special-damage claims without first doing extensive exchanges and analyses of documents.

The owner's first line of defense is a lease provision excluding the owner's liability for such consequential damages arising out of either negligence, warranty, or strict liability claims. The owner who has such a clause can be free to offer a friendly accommodation, such as the use of empty office space nearby, but cannot be hit with a huge contract liability for the loss of use of the building.

If that defense is not available, the owner can argue that the state's real property–leasing laws or court precedents do not make such business interruptions a foreseeable element of consequential damages in the absence of a fire, collapse, or other visible damage. Exclusions and disclaimers are a cheap and easy preventive medicine to avoid losses in any later conflicts.

If both of these defenses fail, the owner can argue that the work should not have been interrupted and that absent an official health department evacuation order, the decision to stop work was not justified by medical evidence; the tenant could have kept the workplace open and cooperated in the remedial action instead of evacuating. The tenant's scientific experts looking back to the date of the event will have a hard time proving that, as of that date, the decision to evacuate was reasonable. Even if subsequent evidence supported the assertion, the legal standard is knowledge at the time of the action. At issue will be whether the tenant's decision was driven by panic or by sensible planning. The owner will probably take a tough line against compensating for interruption, saying, in effect, "We don't pay for your panic."

The tenant will probably assert in later lawsuits that it was necessary to evacuate workers to protect them from even worse health effects, but the tenant has to convince a trial judge and jury that on the date of the action, the knowledge of risk was sufficient to justify evacuation. The jury may be asked, years after the actual incident, to decide what a reasonable tenant would have done. If the jury cannot hear the supporting evidence of the tenant's expert witnesses because of the rules of evidence (in the *Daubert* decision, the Supreme Court directed lower courts to screen expert witnesses' scientific opinions), then the owner has a strong bargaining position from which to settle any litigation on favorable terms.

Next, the prudent owner will respond that the amount of damages claimed is excessive, because the tenant did not mitigate damages, or lessen the amount of its loss. For example, the questionable cluster of rooms could have been sealed off and crews called in for repairs. Workers could have temporarily shared office space during the installation of new ventilation. Employees who did computer-related work could have telecommuted from home; a van in a nearby parking lot could have continued providing customer service; the owner might even have offered empty space nearby as temporary quarters for the employees who had been evacuated. State court case precedents on the lessee's right to recover damages against a landlord are likely to require that the tenant had taken reasonable steps to reduce the cost of the temporary solution to the building's problem.

INCENTIVES NOT TO SUE THE BUILDING OWNER

If the building owner accommodates the needs of the tenant with prompt repairs or cleaning, rent offsets or reductions during the period of difficulty, and temporary space movements, the tenant will be far less likely to assert legal claims or to litigate the building problem.

An owner who has positively addressed the tenant's needs and concerns has lessened the likelihood that the tenant will find it advantageous to sue. Suing one's landlord over losses from indoor health complaints in one work area, if the building's location and other attributes are all positive, is probably not in the tenant's best interest. The owner's expression of caring, desire to help find and remedy the cause, and assistance to the tenant's short-term needs will all be matters that alleviate the lawyers' predilection that only a lawsuit can win relief. Pragmatic businesspeople can negotiate some means of satisfaction with lesser transaction costs.

The tenant also is unlikely to sue if its losses were paid by its business insurance for such damages as business interruption. The insurance carrier's receptivity to such claims will make a difference in whether the owner is asked to pay. Pollution exclusion clauses are common in insurance contracts but have not yet been challenged in an indoor health-related insurance claim. No one likes the costs of litigation, especially small-business people; this makes lawsuits less likely, though in the event of a significant insurance liability expense, the carrier might use its subrogation powers and sue for damages in the place of the insured person. As of 1997, no court decisions had been reported in which property damage was based on indoor health issues.

If the repair or remediation has been done, the building owner who knows that the tenant is actively considering litigation will negotiate and might reduce the lease renewal rent to keep the tenant in the cleaned and refurbished space. Giving a short-term bargain to accommodate the tenant's desires is far better than facing the intangible elements of adverse publicity in the local real estate market, such as rumors that your rental building was so bad that it had to be evacuated.

COUNTERCLAIM BY THE BUILDING OWNER

After the sick building charge is made, the building owner may also have grounds to sue the tenant or other occupant of the rented building for defamation and breach of contract. The tenant may have breached its lease terms, and its abandonment of the building in a panic certainly affected the continued peaceful enjoyment of the building by the owner and by other tenants.

If the tenant was the likely cause of the losses, then the owner should directly sue the tenant for all the consequential losses, including diminished value of the building for future rentals. For example, if one tenant's furniture-stripping vats released gases that affected the health of other tenants' employees, the owner who lost other tenants or must defend lawsuits by other tenants should consider eviction as well as damage suits against the furniture firm.

Ultrahazardous activities by a tenant that overwhelmed normal operating systems are another concern for the owner, who may argue that its systems were reasonable and normal, but the superseding cause of the illness was the action of a third party, the nearby tenant whose activities caused harm. The work done during the tenant's occupancy may have involved chemicals to which the complaining workers were particularly sensitive, even though the building was not adversely affecting people before the tenant's changes to that building were instituted.

Somewhat more difficult is the case of a tenant whose passive activities, such as blocking a fresh-air ventilation duct during renovations, caused the injury with no direct pollution activity of its own. Here the claims of causation of indoor health problems by inadequate ventilation, an element of the case that is already hard to prove, becomes more difficult and consumes more time and effort.

In an extreme case, the building owner may find that adverse publicity generated by the tenant who evacuated the building makes it impossible to get others to rent that space. Like a classic haunted-house story, the legend that indoor environmental health concerns "haunt" your office building can frighten the less-knowledgeable employees of a potential subsequent tenant. If that extreme case arises, state court trade libel case precedents might be examined, and a suit claiming the spread of false and derogatory information about the owner's property might be considered. These suits are expensive and the outcomes are doubtful, so cautious consideration of less-publicized options may be appropriate.

Chapter 13

Building Owner's Options to Sue Persons Other than Tenants

James T. O'Reilly

Introduction

The owner who incurs losses and potential legal liabilities from sick building syndrome (SBS) or indoor environmental health claims may wish to pass the economic burden of those losses on to someone else. This chapter addresses the five principal targets of such an owner's claim: architects and designers, air-handling equipment vendors, installers and other service providers, manufacturers of other products used in the building, and previous owners or persons responsible in the chain of real estate ownership. *None of these claims is easy to win.* Our purpose in exploring each is to give the building owner some ideas about the options available in the event that the employee claim turns into a financial loss. Of course, as with other portions of this text, readers should consult with their own legal counsel regarding the law in your particular jurisdiction.

Architects and Designers

Suits brought by the owner of a building against those who designed it have become increasingly more evident in the case law, and architect malpractice or similar allegations are one contributing cause of higher construction costs.

The designer or architect's liability is usually covered in contract terms of engagement between the owner and the architect. If the owner signed in haste without negotiating these terms, then contractual exclusions and limitations of a pro-architect contract may leave little room for the negligence claims. Though it is unlikely that a state statute or appellate court precedent goes much beyond these contractual liabilities, the prudent lawyer for an SBS-related building's owner will study all the ways in which litigation might be presented to a court.

Change orders, the midconstruction design alterations that are less studied than the preconstruction layout and design, are part of the evidence that the owner will examine in detail. If the architect's original ventilation plan was sound but a midconstruction shift of vents and intakes led to the loading dock's diesel exhaust filling the northwest corner of the first floor, the builder will want to find out who ordered the shift and why it was not examined in light of the air effects inside the building.

If no disclaimers or exclusions apply, lawyers for an owner will usually complain that the SBS building did not meet state-of-the-art standards set by professional code authorities or the American Society of Heating, Refrigeration and Air Conditioning Engineers. The standards have been updated since the first allegations of SBS were made, and the literature known to reasonable architects would have shown a competent design team that certain measures should have been taken to avoid problems.

If a claim against an architect goes to trial, the building owner has a difficult set of proof steps. First, expert witness opinion about causation of SBS has all the problems of showing adequate proof of causation. Next, the judge needs to interpret the contract to include a nondisclaimed duty of due professional care. Then the duty of care in the community, the norm of what constitutes malpractice in professional liability suits, will be debated by competing witnesses— would a reasonable designer in this community have taken these precautions? Then the specific act or omission will be examined—did the designer ignore a well-known sign of SBS causation?

The architect or designer's knowledge of uses is especially crucial in cases where the building was rehabilitated or reconstructed from an original, perhaps historic, design. The renovator is less likely to be found negligent than the designer of an entirely new building.

In cases where the architect, designer, ventilation subcontractor, piping and vent installer, and so on are each sued, the question of who should have known about the problem is accompanied by the question of what duty each of the individual defendants had to catch and correct others' mistakes. If the claims are of secondhand tobacco smoke, off-gassing of wall coverings, intake of pollutants, or the like, then a great deal of second-guessing will take place at this stage of the case.

A final note on liability is that although the architect's finished building may look like a product, from a legal standpoint the architect has supplied a service. Court decisions do not usually treat an architect in the same manner as a maker of a product. Especially when a customized design is involved, this distinction makes it more difficult for the owner to recover damages against the designer.

Air-Handling System Manufacturers

This section addresses the owner's possible legal claims against the supplier of air-handling equipment after sustaining losses from a tenant's SBS claim.

The liability of a manufacturer of heating, ventilation, and air-conditioning (HVAC) products differs from that of architectural and design services in that a manufacturer has more exposure to legal liability if (and *only* if) the equipment proximately caused the illness or injury. Designers who provide a professional service rather than a product are not held to strict liability; this theory allows a person suing a product manufacturer to win damages for personal injury if the product was defective and was unreasonably dangerous, whether or not any fault of the manufacturer of that product is established. To the extent the cause of illness or injury was found to be a product, strict liability for resulting injuries will be claimed to apply under the law of virtually all the states. This section addresses air-handling equipment.

An owner's first possible claim is for negligence in the manufacture of air-handling equipment. The assertion is that the manufacturer owed a duty to design machines that would not spread illness-causing contaminants or bacteria; that the duty was breached through the fault of the manufacturer, by a design of equipment that failed to meet the reasonable standard of care of similar equipment makers; and that the economic or physical injury (lost work time, lost rental income, and medical care costs of the tenants' employees) was proximately caused by the equipment.

Such negligence cases are virtually impossible to win where a scientific cause-and-effect relationship between machine and harm is not well accepted. So the last element, that of proximate causation of illness from equipment, is a tough barrier.

The proof of machine defects that breached a duty of due care is based on a comparison of the equipment in the SBS building with other machines in the industry. The proof of design failures is established by fixing some objective, widely accepted norm as the standard for what reasonably designed machines are designed to do, and then demonstrating a failure to meet that norm. This showing is made more difficult in indoor air cases by the lack of a universally recognized health-based quality level for indoor air. Older buildings with machines dating from periods before current standards took effect are, of course, not held to modern standards. And if the equipment was custom-made for the location, with an agreed-upon set of specifications for the site, the contract probably disclaims liability; even if it does not, the norms of reasonable machine design do not apply.

The fact that the equipment maker is held to strict liability adds to the plaintiff's chances of recovering damages. Strict liability is most useful for the set of proofs of wrongful assembly or manufacture, for example, a defect of a missing machine part that should have been inserted in the factory and that causes mold and mildew to concentrate in the ductwork and grow to potentially harmful levels. If the manufacturer intended for the part to be there but made mistakes that left it out, then strict liability with no proof of fault may be imposed on the manufacturer. Such cases rarely go to trial on issues of fault, since the manufacturer did not intend for a machine to leave the factory with a part missing.

Owners should not expect to use strict liability for multimachine systems; because strict liability was intended to improve quality control at the manufacturing end, an injured person will not be able to assert strict liability for a defect that did not exist at the time the product left the control of its manufacturer. Factors such as installer errors, wrong-way connections, and erroneous cross-connections with other machines excuse the product from strict liability and require the building owner to prove negligence on the part of the machine manufacturer. The modification of a system after it left the factory also extinguishes the chance of winning a strict liability claim, because that intervening act reduces the manufacturer's legal duties. And so-called machine defects are sometimes a result of poor maintenance or inadequate operator training.

When the Supreme Court in the *Daubert* decision told lower courts to examine expert opinion witnesses before trial to screen out speculation and poorly supported reasoning, the result was closer attention by trial judges to the level of support that a proposed expert witness can cite for his or her testimony. The indoor environmental quality (IEQ) expertise of the plaintiff's witness was discussed in Chapter 10. A witness with expertise regarding the HVAC contribution to illness is especially hard to find, for such a person must be both an industrial-hygiene specialist and a professionally experienced building-maintenance expert to be able to connect the HVAC system to claims of unsafe working conditions. But rebuttal witnesses are likely to include operating engineers who run systems for comparable buildings, ventilation-engineering specialists, custom-design HVAC vendors, and health professionals. It is easier to cast doubt on one of these claims than it is for a claimant to prevail with a preponderance of evidence.

Service Contractors and Installers

Potential lawsuit defendants who installed products into the building include those who assembled and installed such items as ductwork, machinery, carpets, and wall coverings as a service provider. The previous section addressed the liabilities of the makers of the installed products.

Complaints related to odor and perceived change in the workplace atmosphere often come as a result of a new installation; for instance, complaints occur on the Monday after a weekend carpet change or a Friday repainting of office walls. A few of the complaints may rise to the level of seriousness that warrants investigation by a safety professional with appropriate measuring and air-sampling equipment. The investigating safety professional is asked to determine whether odors occurred that were more than the routine, transient odors that commonly accompany interior painting, wallpapering, carpet installation, and the like.

The safety professional's first task is usually to find out what changed and what effect the change had on occupants. In the area immediately around the new installation or in the area whose air is fed by new ductwork and air-handling equipment, the sampling should test for indications that there was something more than the normal indoor odors of new paint, wallpaper, rugs, and so on. Expectations of the levels of odors vary, of course, and the difficulty of an SBS case is that the complaining occupants' individual susceptibilities are so variable that the acceptable level of a "normal" transient odor is anyone's guess.

Ideally, the safety professional will find that the installation changed the baseline condition of the indoor air in a measurable way and that the measurable change dissipated over a certain number of hours, leading to a lessened frequency of complaints. In practice, the size of the installation, the nature of the chemical or equipment involved, and other physical factors form one part of the equation; but the missing piece is a preinstallation measurement that can serve as a background level. Few, if any, buildings perform routine measurements of air quality before installing new carpet, paint, equipment, and the like, so the safety professional cannot compare Day 2's air sample to Day 0's.

Objectively proving that the installer caused SBS is virtually impossible without some background level against which the postinstallation workplace environment can be compared. Laboratory measurement of the carpet sample, the adhesive, or the wallpaper or paint can provide well-informed speculation, but it remains uncertain.

If the installer is sued for having contributed to the SBS allegation, the installer's lawyer will press for a more precise definition of the specific contribution. Several defenses can be raised by installers. Perhaps the chemical content of the installed product was not the choice of the installer. Or if ductwork was even arguably a cause of mold and dampness in which biological contaminants grew, the duct installer will attempt to pass the liability back to the building-design team that delivered the blueprints the installer followed. Since installers have the least amount of control over building conditions, being able to change neither the manufactured product nor the ventilation system of the space in which the equipment is installed, juries are perhaps least likely to assign liability to the installer.

Prudent installers include disclaimer clauses in their contracts. Sophisticated painting contractors include standard terms that require the building owner to provide adequate ventilation and that exclude legal responsibility for the indoor air effects of the paint, wall covering, and so on.

If the painting contractor supplied the paint, the duct installer supplied the equipment, and the other installers also wrapped product and service into one price, the lawyer for the building owner should consider separate counts of negligence as to the installation and strict liability as to the product. This is simply an arrangement of the court papers and does not by itself increase the potential that the owner will recover damages.

PRODUCT MANUFACTURERS

This section relates to liability of the manufacturers of wall coverings, ductwork, flooring and carpet, and the like and explains how these companies are potential defendants in suits brought by the building owner regarding SBS allegations.

First, the building owner needs evidence of the actual materials that are alleged to have caused harm. A very practical caution is required: In the rush to address urgent complaints by workers asserting air contamination, the janitorial service or repair persons may dump all of the offending ceiling tile, rugs, and so on into the trash and discard it. Empty cans of solvent, adhesive, or paint may be thrown out as part of the cleanup; their departure diminishes the chance for proving any future liability lawsuit. Samples should be preserved and a written description of the location from which they were removed should be added on an identification tag.

The second piece of evidence the owner needs is the written documentation that accompanied the order. If a product was described in the order by brand name and model number, or if a brochure or (in rare cases) a material safety data sheet (MSDS) was attached, the safety professional can evaluate the installed material with greater specificity.

If the model number and product manufacturer are identified, the safety professional can obtain the correct MSDS. This is a multipage summary of precautions that can help avoid injuries when the chemical or product is used. For example, an MSDS for carpet adhesive may caution, "Use only in a well-ventilated area." It lists hazardous chemical compounds in the finished product, describes any government standard that applies, such as that of flammable liquid, and describes the symptoms of problems that may occur with the substance in the industrial manufacturing environment, for example, whether daily workplace exposure for X years has been reported to increase the occurrence of liver cancers.

A word of caution: the MSDS is *not* a catalog of what can be wrong with office workers who smell the chemical or mixture during an office installation. It is aimed at the worker who handles the chemical constantly in a manufacturing setting. Lay persons who read the sheets for commonly used materials could mistakenly believe that life is a series of dangerous chemical exposures. The sheets do not describe the consequences of everyday common exposure to the chemical; rather, they focus on effects resulting from extended manufacturing exposures that inundate the user with splashes, smells, and skin exposures.

Once the facts about exposure to the product are available, the owner's legal counsel will evaluate how likely it is that a jury will award damages against the manufacturer. As with earlier sections, negligence or strict liability are the probable legal theories that apply here. A separate grounds for suit, breach of warranty, is less likely since most vendors use language in their sales agreement that disclaims any implied warranties.

Economic losses alone cannot be recovered by a strict liability suit; only personal-injury claimants can bring such cases. The affected individual might argue that a defect in the paint or carpet released a harmful chemical vapor that was unreasonably dangerous in its risk, beyond any benefit of the product, and that the unreasonably dangerous feature had proximately caused the injury. These individual suits will be vigorously contested; the plaintiff worker must show that there was a sale of a product, that the product when used was in the condition in which it left the control of the manufacturer, that the risk to users of a particular chemical constituent exceeded its benefit, and that the chemical exposure had proximately caused the specific harm to this worker. Asbestos, an insulating material that has caused a particular disease with specific patterns of lung effects, is an example of a chemical for which individual workers have successfully asserted strict liability. Rock containing radon-emitting particles is a product but its use in a building's foundation is unlikely to encounter the same sets of successful individual suits.

Negligence is the legal theory under which owners try to recover economic losses from a manufacturer of an installed product, if they sue at all. Duty, breach of duty, proximate causation, and actual injury must be proven. Intervening causes and the negligent acts of others may supersede the manufacturer's liability.

A manufacturer's duty toward foreseeable users includes making a reasonably safe product for indoor use when the product is used according to its label instructions. At trial, the judge will decide whether a duty of care existed. If the label of the newly installed product provides directions for ventilation, methods of installation, adhesives to be used, and so on, then the manufacturer will look closely during the pretrial phase of any lawsuit at whether the product was installed as directed. Negligence duties are specific to the role that each party plays in getting the product to the end user. A wall-covering manufacturer who warns against the use of adhesive X is not going to be held liable in a negligence case if adhesive X is applied and the smell of X sickens nearby workers. This would be a case of superseding cause (installer error supersedes any effects attributable to the carpet maker), and the installer could be held liable for negligence.

Breach of the duty of care by the manufacturer, the key element of liability based upon fault, arises if a reasonable manufacturer would have designed or formulated this product differently and the difference would have had a safety or health effect on the workers. If the duty is set at a level related to the work of a reasonable manufacturer, then the unreasonable choices and unreasonable actions of the defendant company would be alleged as the basis for damages.

Expert testimony is needed concerning the actions that a reasonable manufacturer would have taken with this product and how this defendant breached its duty. For example, a comparison might be drawn among five brands of paint. The only one to contain ingredient Z was used on the wall in question, and the medical literature shows that allergies to Z are prevalent among persons like the plaintiff. So the plaintiff's lawyer will seek out a health expert familiar with paint who can testify that the health effect of Z was foreseeable, that use of Z on indoor installations was unreasonable, and that Z-containing paint would not have been sold by a reasonable manufacturer (or would have been sold only with strong cautionary warning statements).

Breach of duty claims are hardest to support where an *omission* rather than an action is claimed. The claim is that the defendant company took no protective action while other manufacturers had done so. For example (and not as a judgment on cause), the claim could be made that a building's air-conditioning machinery was not designed to introduce a chemical preservative that kills bacteria in the water used in the cooling tower, and the absence of such a biocide allowed more biological contamination than other, properly designed equipment with biocides would have permitted. Proving that an omission or failure to act allowed an injury to occur is less convincing to an average juror than proving that a manufacturer's addition of a chemical caused harm among those who breathed large doses of it.

If the duty existed and was breached, the lawyer seeking to pin liability on the product manufacturer is then required to show proof of proximate causation. The witnesses will prove that the product (without any other intervening cause) was more likely than not to have caused this physical injury. The plaintiff will need medical records, publications from the scientific literature, and expert testimony. The expert testimony must be more than speculation and the expert must be sufficiently qualified to survive pretrial screening for reliability.

Connecting the injury to the product is the last of the elements of proving negligence. The medical connection between the product and the kind of injury claimed might be speculative and the plaintiff may be relying on clinical ecology, a creative but not widely accepted view of diagnosing environmental health effects from low levels of chemical exposure. Chapter 4 discussed the medical issues in greater depth.

From the product manufacturer's viewpoint, lawsuits that allege a connection between product exposure and injury are a threat to survival, not just a litigation nuisance.

REAL ESTATE AND/OR BUILDINGS' PRIOR OWNERS

Sometimes, the current building owner who conducts an SBS inquiry discovers that a condition with potential effects on employee health had been known by previous owners. "We've always had problems in that area of the building" is a message that a relatively new owner hates to hear.

A building that failed a radon test is a simple example; concealing the failure and failing to take remedial measures is a serious breach of the responsibilities of the property seller. Fraudulent concealment is a potential basis for a claim against the seller and perhaps its agent as well, depending on the extent of knowledge and conspiracy to conceal the known defect. The more remote or esoteric the claim of injury, the less likely it is that buyers will ask these SBS-related questions that would reveal the problem.

Buyers usually are held to a reasonable duty to inspect the property before purchase. If the claim of concealment is made, the jury must decide first what the reasonable buyer would have done and then whether the actions of this buyer were reasonable. Unless some concealed defect existed, the buyer should reasonably have found the obvious dangers.

Where does the pattern of past air-quality problems fit in a liability claim against a previous owner of the building? Does a building seller have a legal duty to describe to the owner the occurrence of past claims of sickness caused by the building? Can the buyer be expected to know the conditions of possible microbial contamination that lurk within a damp seventh-floor air-return duct? These questions have no clear legal answer today; they are among the issues that will no doubt be debated in the twenty-first century.

If we may speculate on legal evolution, it is probable, but not certain, that SBS claims will increase, and the level of care of the reasonable and prudent buyer will increase as well. Courts will probably consider that, as SBS becomes more widely known in the commercial real estate community, the reasonable buyer should ask about the track record of indoor air quality for this building. More courts will expect careful questions from buyers. Sophisticated building managers with low complaint rates are likely to retain records of complaints and service calls, which the buyer's assessment team can use as a basis for determining the early-warning signs of building-illness claims.

We can predict a trend among prepurchase assessors. Property buyers who routinely examine soil and groundwater Phase I and Phase II evaluations, watching out for Superfund and hazardous waste cleanup liabilities, will probably grow more sophisticated in asking the seller about indoor conditions. If this occurs, then an indoor equivalent of the widely practiced environmental screening will occur, a sort of indoor Phase I checkup. As with the exterior real estate examination known as a Phase I survey, the indoor examination will include researching the past uses of the building; noting the presence of laboratories, paint stores, large print shops, dry cleaners, and the like in an office or mixed-use building; noting the age of the building for purposes of lead paint, asbestos, and so on; finding residues of past dusts from metalworking in the case of a renovated factory-type structure; and other relevant details that may signal a change from previous uses.

Chapter 14

Insurance Law Implications of Indoor Health Claims

JAMES T. O'REILLY

NOTICE REQUIREMENTS

Sick building syndrome (SBS) and related claims can be expensive to manage and to settle, especially in the case of litigation. Insurance carriers should be made aware of litigation as soon as possible after the employer and/or building owner are sued. Policy terms typically require prompt notice so that the field adjusters and lawyers for the carrier can prepare an estimate of the liability reserves needed to deal with potential losses.

Early notification of the potential claim by the policyholder is important to the carriers' ability to later defend the claims. Failing to give notice may jeopardize the defendant company's insurance coverage and may result in denial of any liability payments by the insurance carrier.

EXCLUSIONS OF POLLUTION COVERAGE

The majority of commercial insurance policies follow the standard format, used since the mid-1980s, of excluding coverage for all of the effects of pollution. When an SBS-related claim arises, the carrier has a strong economic interest to deny coverage on the basis that the claim involves pollution. This has been a common practice in Superfund site cleanup cases, triggering lawsuits, called coverage suits, brought by the company that is stuck with the cleanup costs.

At the time this book went to press the reported cases involving denial of coverage for indoor air quality liability had not yet been litigated; however, it is predicted that carriers will disclaim any duty to pay because the injuries were caused by pollution that is excluded by the so-called absolute pollution exclusion.

The pollution exclusion clause does not squarely address indoor health issues. But since asbestos, PVC dust, and other materials have been held in court to be pollutants, it is very likely that insurance claims for indoor air hazards will also be subjected to the exclusion of pollution events.

Insurance-coverage litigation has been a dramatically developing area in environmental-waste cases and might become significant in SBS cases in the future. Denials of coverage will arise more frequently than with, for example, claims of fire or flood damage that interrupts the workplace. Cautious phrasing in correspondence about the loss is critically important. Acknowledging the claim as a "pollution" assertion in a letter by the insured company, to the insurance carrier, would be the worst possible error. An insurance professional and an attorney skilled in litigating disputed coverage cases should advise the building owner early in the process.

Duty to Assist the Insurance Carrier

The insured party, typically the building owner, should offer to assist the claims adjusters and consultants who work with the insurance carrier. Factual information about the building conditions should be shared. Disputes about the legal interpretation of coverage are not likely to be affected by the actual cooperation between the owner and the insurance carrier's representatives.

Other Insurance Issues

The coverage of several past policies may have to be invoked if the current insurance carrier denies liability, on the basis that the adverse event occurred before the current policy and carrier came on the scene. In some cases, a consultant familiar with insurance policy archaeology might be hired to determine which carriers insured the building at the time the dangerous conditions arose.

More frequently, the extent of the alleged harm to workers was manifested only in a present-day climate of publicity about sicknesses associated with indoor air environmental conditions, so the previous set of insurance policies would not be likely to afford coverage.

Chapter *15*

Case Study: The Brigham
and Women's Hospital Experience

Tamara Lee Ricciardone, Esq.

Introduction

Brigham and Women's Hospital (usually referred to as the Brigham), a major
Boston teaching hospital, began experiencing an increase in air-quality complaints by its employees during the three-month period of March through May
1993. Employees in the operating room suites reported symptoms that primarily
involved irritation of the skin and upper respiratory tract and, in some cases,
asthmalike bronchoconstriction. The Brigham conducted an investigation of its
workplace environment and concluded that the operating rooms' ventilation system required modification. In addition, some sterilization and cleaning agents, as
well as latex gloves, were identified as a potential cause of allergic responses in
some employees, although no industrial-hygiene measurements exceeded the applicable safety standards. Significant improvements were made to the ventilation
system. The hospital reduced its use of natural rubber latex products and the offending sterilization and cleaning agents.

In late 1993, the Brigham experienced additional complaints by employees
working in other areas of the hospital, predominantly on the upper floors of the
Patient Tower building. Similar to those in the operating room suites, these complaints included upper-respiratory symptoms and skin rashes. However, the picture became more complicated when some employees began to claim that they
had developed multiple chemical sensitivities (MCS), a diagnosis that the medical community widely disputes as characterizing an actual disease condition. The
hospital again investigated the workplace environment and made improvements
to the ventilation system and the procedures for medication use; a number of

substances containing chemicals that were identified as causing reactions in some employees were removed. Once again, although some substances were identified as potential irritants, no industrial-hygiene measurements indicated that these substances exceeded the appropriate indoor environmental quality (IEQ) standards.

This chapter explores the initial complaints made by employees at the hospital and the subsequent investigation the hospital conducted of the ventilation systems and the industrial hygiene measures employed. The ventilation system the hospital uses and the improvements made following its investigation are also described. The hospital conducted industrial-hygiene sampling for various chemicals and biological contaminants that may have been affecting IEQ, and the results of that testing are examined. Finally, the legal aspects of the Brigham's experience are addressed and insights and recommendations for the handling of environmental complaints are discussed.

BACKGROUND

Brigham and Women's Hospital is a 751-bed, nonprofit teaching hospital affiliated with the Harvard Medical School. The Brigham employs a staff of over 8,000 that is dedicated to patient care, medical research, and the training and education of health care professionals. The Brigham's roots go back to 1832 with the founding of the Boston Lying-in Hospital, which in 1896 merged with the Free Hospital for Women (founded in 1875) to become the Boston Hospital for Women. In 1976, a merger of the Boston Hospital for Women, the Peter Bent Brigham Hospital (founded in 1913), and the Robert Breck Brigham Hospital (founded in 1914) created the Brigham and Women's Hospital. In 1975, ground was broken for the new building that would become the largest hospital construction project in the history of Massachusetts. In 1980, the Brigham and Women's Hospital opened its new Patient Tower, a 16-story, 712-bed facility. Most recently, the Brigham affiliated with Massachusetts General Hospital and North Shore Medical Center to form Partners HealthCare System, Inc. Since that time, the Brigham has experienced continued growth. Brigham personnel have won two Nobel prizes, and the Brigham is a nationwide leader in total grants, contracts, fellowships, and funding received from the National Institutes of Health. The hospital provides women's health services, operates New England's largest birthing center, and is a designated regional center for high-risk obstetrics and neonatology. The hospital has also received national recognition for its transplant programs, its cancer research efforts, its performance of joint replacement and orthopedic surgery, and its treatment of arthritis, rheumatic disorders, and cardiovascular disease.

INCREASE IN HEALTH COMPLAINTS

In 1993, the Brigham experienced a steady increase in the number of health complaints among employees working in the operating room suites (OR), which are located on the lower level of the Patient Tower building (Floor L1). In March through May 1993 particularly, the number of health complaints including incident reports and visits to the employee health services increased dramatically. This increase prompted the Brigham to undertake an extensive review of the environmental conditions in its operating rooms, including their ventilation systems. Employees at that time complained of a complex symptomatology that included headaches, fatigue, dizziness, nasal congestion, shortness of breath, wheezing, chest tightness, coughing, throat tightness, eye irritation, rash, itching, and hair loss (the OR symptoms). During the period of April 1, 1992, to June 1, 1993, from a total of 299 nonphysician OR employees, 50 (16.7 percent) complained of OR symptoms. Eight of the 50 employees had OR symptoms limited to the skin, and 42 complained of additional symptoms. Forty of these complaints occurred during the months of March, April, and May 1993. In contrast, during that same period, 17 (2.4 percent) of the hospital's non-OR employees complained of the OR symptoms: 15 of these employees voiced complaints limited to the skin; the remaining 2 employees complained of additional symptoms. Based on these statistics, the Brigham determined that its OR employees were experiencing a seven times greater risk of suffering from the OR symptoms.[1] By the end of October 1993, approximately 180 OR employees reported a variety of symptoms that were potentially related to the air quality and other environmental conditions of the operating rooms.

On April 9, 1993, an employee of the Brigham made an informal complaint to the Occupational Safety and Health Administration (OSHA) and alleged that OR employees were exposed to nitrous oxide leaks; were experiencing asthma, headaches, and chest pain; and were not informed of air-monitoring results.[2] The Brigham launched a full-scale investigation into the operating room environment. This investigation was conducted as a joint effort among the departments of nurse management, employee health services, environmental health and safety, engineering, human resources, and facilities management. In addition, in June 1993 the Brigham entered into a contract with an environmental-consulting firm, Environmental Health and Engineering of Newton, Massachusetts (EH&E), to conduct industrial-hygiene sampling and to review the OR environment.

---

[1] Howard Hu, M.D., *Draft Report on Health Complaints and Environmental Exposures in the Brigham and Women's Hospital Operating Rooms,* June 16, 1993.

[2] The Hospital received no formal citation from OSHA following this complaint.

INVESTIGATION OF THE OPERATING ROOM SUITES [3]

Hired by the Brigham to provide scientific and engineering consultation services regarding the recent symptoms that the OR staff reported, EH&E began its investigation in June 1993 by undertaking a comprehensive review of the OR environment. This review included evaluating the OR air quality through the use of continuous monitors and random sampling of suspect agents. EH&E also evaluated the general and local ventilation systems servicing the OR.

The OR is located on the L1 level of the Patient Tower and contains 32 operating rooms. The area consists of approximately 20,000 square feet (sq.ft.). Some of these facilities were built during the original construction in 1980. Operating rooms 1 through 20 are part of the original construction, while rooms 21 through 25 were constructed in 1984, rooms 26 through 29 in 1986, and rooms 30 through 32 in 1989. EH&E's initial investigation included all of the operating rooms on the L1 level.[4]

Ventilation Systems

The major mechanical heating, ventilating, and air-conditioning (HVAC) system servicing rooms 1 through 25 is a dual unit, separated by a common wall, and provides 52,000 cubic feet per minute (cfm) of supply air (outside air that is pulled into the system to ventilate the building) to the OR areas. The system was designed to offer 100 percent redundancy so that one side could be operating fully while the other is down for servicing. The system ventilates approximately 12,000 sq.ft. of the total OR area. The mechanical system serving rooms 26 through 32 has a nominal capacity of 20,000 cfm. This unit services approximately 5,900 sq.ft. and is located in a penthouse on the rooftop of the adjoining building, the Amory Laboratory[5] (see Figures 15.1 and 15.2).

The supply air for the OR changes 20 times per hour. The system is balanced to maintain the OR at a positive pressure relative to its surroundings. Each room is equipped with its own humidifier system. Ceiling diffusers provide supply air to the room, and filters remove 95 percent of particles sized at 0.3 micrometers

---

[3] The environmental data contained in this chapter is an overview of multiple environmental reports that have been generated during the Brigham experience. For a full and complete understanding of all of the testing, reporting, and findings, one must review each of the specific reports generated. The author does not intend for this chapter to be a complete summary of all environmental testing that has been completed at various stages of the investigations.

[4] Environmental Health and Engineering, *Indoor Air Quality Investigation—L1 Level Operating Room Suites—Executive Summary Final Report,* August 2, 1994, at 25.

[5] Ibid. at 25–26.

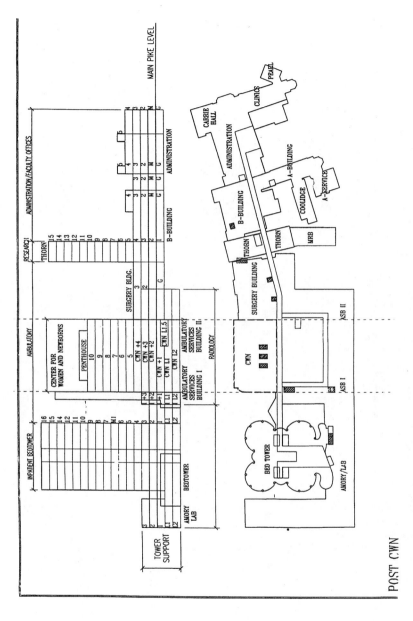

Figure 15.1. Layout of the hospital buildings.

231

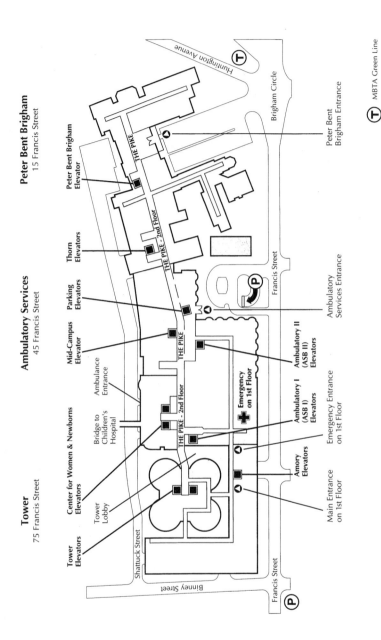

Figure 15.2. Map of hospital grounds.

(μm). The supply air enters the OR with much lower velocities than used in office space design. Return air (the air that is extracted from the room) from the OR enters through return grilles located on the lower walls in the corners of each room. The design of the supply air diffusion/return air extraction system is intended to minimize exposure to particles in the OR.[6]

The unit that serves rooms 1 through 25 draws its fresh air from a common air shaft in the North and South L2 Mechanical Rooms, as do three air-conditioning units on the L2 level of an adjoining building, ASB-I. Through the use of an economizer cycle, the unit operates to vary the amount of outdoor air entering the system based on the thermal conditions of the outdoor air. The return air fans dump their return or relief air into a common relief air shaft. As the economizer cycles demand more outdoor air, the volume of relief air increases.[7]

Following its review of the ventilation systems, EH&E determined that both the outdoor air shaft and the relief air shaft servicing the unit were undersized for maximum demand. This undersizing allowed static pressure to increase as the economizer cycles operated and resulted in pronounced fluctuations in the airflow. These fluctuations adversely affected the relative volume pressurization of the OR. To avoid these problems, the Brigham locked into a fixed position the outdoor air dampers of all the units served by these systems.[8]

In addition, the system servicing rooms 1 through 25 was supposed to humidify them. The humidifiers in many of the operating rooms had been problematic and therefore were not operating at the time of EH&E's initial investigation. EH&E engineering personnel contacted the manufacturer's representative of the humidifiers to develop a solution for their service problems.[9]

The air-handling system that services rooms 26 through 32 is located in a penthouse on the third level of the building. EH&E evaluated the system and found that, under some weather conditions, the air intake did not meet the recommended levels for dilution from the exhaust systems located on the roof. The Brigham worked with EH&E to reconfigure the exhaust outlets and their intakes to correct this problem.[10] Further, in late 1993, the Brigham discovered that the system serving rooms 26 through 32 was leaking air out of some of the ductwork joints, causing approximately 25 to 30 percent of the air coming from the unit to leak before reaching its destination in the OR suites. In addition, the Brigham discovered that sprinkler system pipes installed in the late 1980s blocked some of the filter access doors. Also, after a chilled water spill in the winter of 1994,

---

[6] Ibid.

[7] Ibid. at 26.

[8] Ibid.

[9] Ibid. at 28.

[10] Ibid. at 27–28.

the fresh air grille was relocated from the face of the penthouse due to odor problems. Its original placement on the front of the building allowed odors from the chilled water spill, as well as food odors from the kitchen, to enter the air supply.[11]

EH&E investigated several other potential sources of contamination, including the Cidex (glutaraldehyde) local exhaust hood, the steris sterilizers, Exomat X-ray developers, ethylene oxide sterilizers, and ethylene oxide aeration chambers. Its findings with respect to each of those potential sources include the following: (1) the Cidex hood exhaust was positioned in such a way that it caused leaking of glutaraldehyde fumes into the ceiling above the operating room suites; (2) the steris sterilizers' overall local exhaust ventilation was satisfactory; (3) the Exomat X-ray developers were exhausting fumes into the return air and the flexible ductwork was allowing leakage; and (4) the ethylene oxide sterilizer process ventilation did not completely control emissions, resulting in exhaust fumes from that sterilizer spilling into the air supply. For each of these problems, EH&E designed and commissioned a new ventilation strategy.[12]

EH&E discovered that the outflow for the exhaust shaft servicing such processes as pharmacy radiological hoods, steris sterilizers, and Exomat emissions for X-ray development is located close to the supply-air shaft. The relationship between the two shafts is such that air can migrate from the positively pressured exhaust shaft to the negatively pressured outdoor supply-air intake shaft under certain conditions. Tracer tests indicated that up to 5 percent of the exhaust air might migrate into the supply outdoor air shaft. This migration did not necessarily indicate a dangerous situation. The nature and concentration of the contaminants in the exhaust air, as well as the quantity and quality of supply air, determine the extent of a potential hazard. The tracer test indicated that a pathway existed for the exhaust air to migrate into the supply air, probably due to small perforations in the concrete common wall between the two air supplies and the pressure relationships between them.[13]

EH&E identified a number of ways in which the design and operation of the HVAC system could be improved, and the Brigham took appropriate actions to correct these problems. These actions included relocating the exhaust fans for the exhaust shaft outside the building to ensure that negative pressure was maintained within the exhaust-air supply and converting the OR to a constant-volume

---

[11] Deposition of Robert A. Murray, taken on August 18, 1995.

[12] Environmental Health and Engineering, *Indoor Air Quality Investigation—L1 Level Operating Room Suites—Executive Summary Final Report,* August 2, 1994, at 29; Environmental Health & Engineering, *Letter to Robert Murray Regarding Results of Initial OR Suite Investigations and Actions Taken to Date (EH&E 93.152),* August 16, 1993, at 2–3.

[13] Environmental Health and Engineering, *Indoor Air Quality Investigation—L1 Level Operating Room Suites—Executive Summary Final Report,* August 2, 1994, at 27.

system in order to maintain a constant positive pressure in the fresh-air supply shaft. The Brigham also improved the local exhaust control within the interior of the OR areas to ensure that a larger volume of fresh air was being supplied to the area. The establishment of a more appropriate positive to negative pressure relationship between the supply and exhaust shafts minimized the potential for chemicals to enter the OR areas through the ventilation systems.[14]

The Brigham engineering personnel completed the refurbishing of the HVAC system, including humidifiers, in rooms 1 through 14 in the late fall of 1994. The scope of the renovations was extended to include rooms 21 through 31.[15] Engineering personnel removed obstructions, improved the humidifiers, and added additional safeguards to the humidifiers to control the injection of steam into the duct system in which there is no airflow. Monitoring of the air quality in the operating rooms continues to occur, and reports indicate that the general indoor air quality (IAQ) in the OR area is satisfactory.

Industrial Hygiene Sampling

The Brigham was concerned that certain chemicals used in the OR might be affecting employees' health. Early on, employee health services confirmed that a number of employees reporting symptoms had allergic responses to latex as determined by a radioallergosorbent serum test (RAST).[16] Other chemical exposures were also a concern. EH&E conducted extensive air sampling during operational hours from June 1993 to December 1993 and obtained the following results:

• *Formaldehyde.* Formalin, containing formaldehyde, is used in the OR as a tissue preservative. Formaldehyde levels were generally less than 1/100th of the occupational health standard of 750 parts per billion (ppb) for an eight-hour time-weighted average. The World Health Organization recommends that formaldehyde levels not exceed 80 ppb for a 30-minute average to prevent IAQ complaints from sensitive individuals. The American Society of Heating, Refrigerating and Air-Conditioning Engineers (ASHRAE) recommends that formaldehyde levels in indoor air be less than 1/10th of the applicable threshold limit value. The measured results were well below these criteria for IAQ with the exception of one sample collected in the pathology room that EH&E stated did not

---

[14] *Occupational Safety and Health Administration Report,* January 14, 1994.

[15] Environmental Health and Engineering, *Letter to Robert Murray Regarding Humidification of Brigham & Women's Operating Rooms (EH&E 94.034),* September 22, 1994, at 1–2.

[16] Environmental Health and Engineering, *Indoor Air Quality Investigation—L1 Level Operating Room Suites—Executive Summary Final Report,* August 2, 1994, at 15.

represent the general room concentration as it was measured at the dispensing area during the use of a formaldehyde solution.

Some employees expressed concern regarding their exposure to formaldehyde absorbed in the dust generated by the use of disposable paper products and formaldehyde-containing scrub coats. EH&E evaluated the formaldehyde exposure based on the highest expected dust concentration in the OR environment. EH&E determined that personnel exposure to absorbed formaldehyde in the dust from the paper products and garments represented approximately 1/200th of the exposure to the lowest median concentration of gas-based formaldehyde. Consequently, airborne exposures to formaldehyde through this mechanism was insignificant.[17]

• *Glutaraldehyde.* Glutaraldehyde is a germicide used to disinfect instruments. The occupational health standard for glutaraldehyde is 200 ppb, and the levels in the OR were mostly below the limit of detection (i.e., not detectable) and well below the occupational health standards. The results ranged from .066 ppb to 6.6 ppb, the latter result indicating levels during the use of glutaraldehyde in the anesthesia technician workroom. Once the Brigham removed glutaraldehyde from the anesthesia technician workroom, the highest recorded level was 1.3 ppb, which is 1/150th of the occupational standard of 200 ppb. Although these readings were well below the appropriate standards, the Brigham discontinued use of glutaraldehyde in the OR environment with one exception: glutaraldehyde is still present in the fixative used for X-ray film development.[18]

• *Volatile organic compounds.* Volatile organic compounds (VOCs) were measured in the OR during surgical procedures to evaluate the atmosphere under operating conditions. All of the VOC concentrations measured were well below the recommended values. The levels of the most predominant compounds were 1/10th (isoflurane) to 1/3000ths (acetone) of the recommended levels for satisfactory IAQ.[19]

• *Anesthetic gases.* In addition to the VOC analysis, EH&E also measured the levels of anesthetic gases. Again, EH&E collected samples during normal operating procedures. Four primary anesthetics are used at Brigham and Women's: nitrous oxide, enflurane, isoflurane, and desflurane. Enflurane was generally not

---

[17] Environmental Health and Engineering, *Indoor Air Quality Investigation—L1 Level Operating Room Suites—Executive Summary Final Report,* August 2, 1994, at 7–9.

[18] Ibid. at 9–10; George Weinert, CIH, Director, Environmental Safety and Health, *Memorandum Regarding Industrial Hygiene Sampling Results for Anesthesia TEE Cidex Room,* December 8, 1994.

[19] Environmental Health and Engineering, *Indoor Air Quality Investigation—L1 Level Operating Room Suites—Executive Summary Final Report,* August 2, 1994, at 10–13.

detectable or detectable only in trace quantities, and isoflurane levels were well below the recommended level of 35 ppb. Nitrous oxide levels were less than the detection limit of 1 part per million (ppm) except for one sample in room 32 that measured 13 ppm. This level was also below both the recommended limits of 50 ppm established by the American Conference of Governmental and Industrial Hygienists and 25 ppm established by the National Institute of Occupational Safety and Health (NIOSH).[20]

- *Latex particles.* No generally accepted occupational standards govern the amount of airborne latex that is acceptable in a work environment. Best practices dictate that, to reduce reactions to this material, levels of airborne latex should be decreased to the lowest feasible levels. Since no standard is available to govern the appropriate levels of latex, the Hospital decided, with EH&E's guidance, to establish its own internal action level of 10 nanograms per cubic meter (ng/m^3). Generally, the Brigham believed that the airborne latex particles in the OR environment resulted from the frequent use of latex-containing rubber gloves, because latex particles become airborne during the use of these. The employee health department recognized that the use of latex gloves was a potential problem in the operating room, and the hospital instituted a substitution program to remove latex-containing gloves from the OR and to replace them with low-latex-allergen products. For most of the air samples measuring latex, the results were at or near the detection limit, indicating that this reduction effort was successful.[21]

- *Ethylene oxide.* The Brigham uses ethylene oxide in the central supply to sterilize surgical tools and equipment. Although no employee exposures were detected, the control equipment in place to capture ethylene oxide emissions allowed some leakage, and as a result the hospital overhauled the local exhaust ventilation system. After the Brigham improved the ventilation system, the air quality was tested. All the sampling results outside the ethylene oxide ventilation plenum showed concentrations well below OSHA regulatory and industry limits. In addition, most of the samples were below the limit of detection, demonstrating the success of the ventilation upgrade.[22]

Some hospital OR staff members expressed concern that employees were being exposed to ethylene oxide through the use of sterilized surgical packs. These packs include paper gowns, cloth, and various surgical tools used in the OR. The employees were concerned that, upon opening these packs, they would be ex-

---

[20] Ibid. at 13–14.

[21] Ibid. at 14–15.

[22] Ibid. at 16–17; Environmental Health and Engineering, *Letter to Robert Murray Regarding Commissioning of the EtO Ventilation System (EH&E 93–152),* January 5, 1994, at 1–2.

posed to ethylene oxide. In response to these concerns, the Brigham investigated the emissions from these packs and found that all the samples collected contained concentrations that were below the detection limit and below the OSHA standard of 1 ppm for an eight-hour time-weighted average. Personnel exposure to ethylene oxide from the opening of the packs was judged insignificant.[23]

• *Acid gases.* EH&E conducted sampling for gases of four inorganic acids: sulfuric acid, hydrochloric acid, hydrofluoric acid, and nitric acid near the acid neutralization system located in the L2 Mechanical Room. The ambient levels of hydrochloric acid, hydrofluoric acid, and nitric acid were all below the limit of detection at all sample locations. Sulfuric acid was present in detectable quantities in two or three sampling locations, and these measured concentrations were determined to be well below the occupational exposure limits.[24]

• *Bioaerosols.* Bioaerosols consist of airborne particles comprising viruses, bacteria, protozoa, pollen, fungi, arthropods, or animals.[25] The concentration of airborne fungi in the OR at the Brigham was extremely low. Only one of the indoor samples resulted in any growth at all, and that sample yielded only 12 colony forming units per cubic meter (cfu/m^3). This concentration was in contrast to the 318 cfu/m^3 present in the outdoor air on the same day. EH&E collected no airborne environmental bacteria in either the indoor or the outdoor samples. EH&E detected a moderate amount of fungi in the bulk sample collected from the instrument room. These fungi consisted of unidentifiable *Mycelia sterilia* and *penicillium,* both of which are commonly found indoors and contain allergenic properties. However, during the air sampling, EH&E collected no *penicillium.* The bulk sample came from grease or other lubricant found on an exhaust grate.[26]

• *Particulates.* EH&E conducted sampling from the OR and determined that the operating room was uncontaminated by particulates. EH&E concluded that the high-efficiency particulate absolute (HEPA) filtration system and positive pressure relative to the surrounding spaces were functioning properly. The spaces surrounding the OR had a higher concentration of airborne particulates, but no particles were foreign to normal indoor environments.[27]

---

[23] Environmental Health and Engineering, *Indoor Air Quality Investigation—L1 Level Operating Room Suites—Executive Summary Final Report,* August 2, 1994, at 16–18.

[24] Ibid. at 18–19.

[25] Environmental Health and Engineering, *Environmental Measurements Reference Sheet,* August 25, 1994, at 9.

[26] Environmental Health and Engineering, *Indoor Air Quality Investigation—L1 Level Operating Room Suites—Executive Summary Final Report,* August 2, 1994, at 20–22.

[27] Ibid. at 22–24.

Corrective Measures

Aside from making the previously described changes to the ventilation system, the Brigham took the following actions in response to the investigation of the OR:

- The operating rooms switched to a low-latex-allergen glove.
- The Brigham took the Cidex hood off-line and significantly reduced the use of glutaraldehyde.
- The Brigham took the Exomat developer off-line.
- The Brigham began keeping at least one operating room empty so that procedures could be shifted to that room in the event of a complaint of poor indoor environmental quality (IEQ).
- The operating rooms were power cleaned, a process that required closing the OR for several days over the Thanksgiving weekend in 1993. Power cleaning included intensive cleaning above the ceiling and replacing all ceiling tiles; cleaning all air-supply and exhaust ducts; breaking down and cleaning all medical and electrical equipment; stripping, washing, and re-finishing all floor areas; and reconfiguring the space for better equipment storage.[28]

INVESTIGATION OF THE PATIENT TOWER

In the fall of 1993, the Brigham experienced increased complaints from employees working in the Patient Tower, Floors 1 through 16 (the Tower). Some employees reported symptoms of fatigue, dry scratchy throat, rash, chest tightness, and shortness of breath. In addition, employees complained that certain odors were causing their symptoms to worsen. EH&E undertook a comprehensive investigation of each of the floors about which employees had complained. The investigation revealed three areas of potential concern: the building-ventilation systems, industrial-hygiene measurements, and medication use. The ventilation system was found to meet applicable codes and be functioning properly. EH&E identified some improvements that could be made to the HVAC systems; however, other than the use of latex gloves, it could not identify any one particular element that was the cause of the employees' ongoing complaints. On each floor of the 16-story Patient Tower, EH&E identified separate and distinct potential causes for employees' complaints. Overall industrial-hygiene sampling demon-

---

[28] Mary S. Fay, R.N., V.P., Nursing, *Letter to Staff Nurses Regarding Workplace Environmental Issues,* October 5, 1995.

strated that the hospital was operating well within the appropriate government standards for good IAQ as well as the more stringent guidelines the Brigham developed for internal use. As no particular agents could be identified as the source of the ongoing complaints, the hospital undertook a general power cleaning of each of the floors in the Patient Tower to eliminate environmental dust and latex allergen.

Ventilation Systems

The Tower is a 16-story high-rise that connects to the adjoining buildings at the first three levels as well as at the fifth. The 400,000-sq.ft.-building is used for patient recovery after surgery or medical procedures. Each floor encloses approximately 25,000 sq.ft. of floor area, of which 10,000 sq.ft. forms a central square core area. The remaining area is divided to form four semicircular patient areas known as the pods. These are referred to as Pods A through D (see Figure 15.3).

The mechanical systems that service the Patient Tower are located on the M-1 level of the building between Floors 6 and 7. All of the HVAC systems for the Tower, as well as some of the fans for the toilet exhausts and general exhaust systems, are located at this level. The system consists of eight air-handler units, AC1-BT through AC8-BT. AC1-BT through AC4-BT are constant-volume systems that provide high-pressure supply air to induction units located around the perimeter of the patient pods. AC5-BT and AC6-BT are variable-air-volume units that provide supply air to the center of the building. AC7-BT provides supply air for nurseries located on Floors 3, 4, and 5. AC8-BT provides supply air to the intensive care units (ICU) on Floors 8, 9, and 10. Only these latter two units are equipped with systems to humidify the supply air.[29]

Each area (except the ICU areas) provided with supply air has an associated return-air fan and two air-conditioning units that are served by a common return-fan system. For systems AC1-BT through AC6-BT, the mechanical equipment room serves as part of the path between the return-air fan and the supply-air fan. Each mechanical room is equipped with spill dampers to relieve excess pressure. These systems have the nominal capacity to supply 335,800 cubic feet per minute (cfm). At minimum conditions, this capacity allows enough outdoor air for two air changes per hour, the recommended minimum amount of outdoor air for patient areas.[30] EH&E noted the potential problem of having the mechanical equipment room function as part of the return-air supply for systems AC1-BT through AC6-BT. Materials generated or released into the mechanical room

---

[29]Environmental Health and Engineering, *Indoor Environmental Quality Investigation at the Patient Tower—Draft,* November 14, 1994, at 45.

[30]Ibid. at 46.

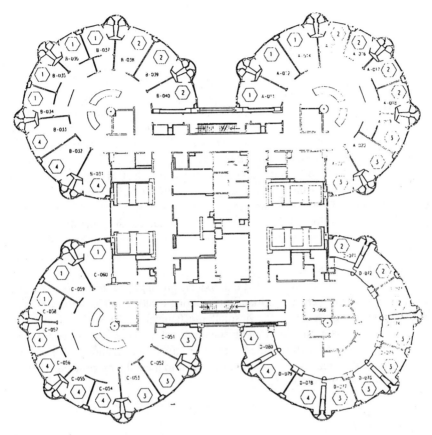

Legend

⟨1⟩   Space Served by AC-1bt

⟨2⟩   Space Served by AC-2bt

⟨3⟩   Space Served by AC-3bt

⟨4⟩   Space Served by AC-4bt

Figure 15.3.   Layout of patient tower.

could be circulated through the occupied areas on the patient floors if the systems were in part using recirculated air during that time.[31]

EH&E's inspection of the HVAC system revealed that each of the air-handling units was operating within its design specifications with the exception

---

[31] Ibid. at 51.

of AC4-BT, which demonstrated an increase in static pressure within the mixing box, reducing the air supply in the system.[32] EH&E further discovered that the outdoor air damper was binding and would not open fully. Engineering personnel corrected the problem, and the system is now working properly.[33]

On Floor 8, an Exomat X-ray developer created odors that migrated into the hallway, and several employees associated their complaints with these odors. EH&E determined that several dampers in the duct system were closed, a condition that created a pressure relationship that allowed odors to migrate into the hallway. Once the dampers were opened, the exhaust airflow improved significantly, the room was rebalanced, and the intensity of the odors diminished.[34]

The HVAC design for Pod 12D is different from that of other areas of the Tower. A small area of the central core is served by AC5-BT. The remaining areas of the pod are served by air-handler unit AC-12-1, which was installed as part of the renovation of Pod 12D. EH&E inspected the operation of unit AC-12-1 and evaluated its performance as normal. EH&E concluded that outdoor air is effectively being delivered to Pod 12D in quantities consistent with the design specifications.[35]

EH&E's investigation revealed that the design and operation of the Tower's ventilation systems were appropriate and met or exceeded all Commonwealth of Massachusetts requirements for hospital mechanical systems. The Commonwealth's requirements are based upon standards set by the American Institute of Architects. However, EH&E noted that under minimum outside air operation, the mechanical systems were not quite sufficient to meet the more stringent standards established by ASHRAE.[36]

Industrial-Hygiene Sampling

Once again, the Brigham was concerned that employees were affected by certain chemicals used in the Tower. EH&E conducted extensive testing of the air quality during operational hours from January 1994 to May 1994. Using a Bruel and Kjaer Model 9652 multipoint, multigas tracer analysis system instrument, which

---

[32] Environmental Health and Engineering, *Patient Tower IAQ Investigation Update (93.371)*, March 24, 1994, at 2.

[33] Environmental Health and Engineering, *Indoor Environmental Quality Investigation at the Patient Tower—Draft*, November 14, 1994, at 48.

[34] Environmental Health and Engineering, *Patient Tower IAQ Investigation Update (93.371)*, March 24, 1994, at 2.

[35] Environmental Health and Engineering, *Indoor Environmental Quality Investigation at the Patient Tower—Draft*, November 14, 1994, at 56–66.

[36] Ibid.

employs a photo-acoustic infrared sensor for detection, EH&E measured the concentrations of carbon monoxide, carbon dioxide, formaldehyde, glutaralde-hyde, and total hydrocarbons. EH&E also measured the concentrations of bioaerosols, VOCs and latex particles.[37] EH&E obtained the following results:

Carbon monoxide. The carbon monoxide concentrations measured at all lo-cations and during all measurement periods were much lower than the ambient air quality standard of 9 ppm. The maximum reading was 4.3 ppm. The indoor levels were consistent with the levels measured in the supply air, suggesting that the minimal presence of carbon monoxide was due primarily to the ambient at-mospheric concentrations. EH&E concluded that carbon monoxide was not a likely contributor to the employee health complaints in the Tower.[38]

Carbon dioxide. EH&E based the carbon dioxide measurements on the 90th percentile readings to prevent skewing of the data due to people breathing in or around the sampling tubes. The average concentrations of carbon dioxide for various areas in the Tower ranged from 448 ppm to 615 ppm. ASHRAE stan-dards call for concentrations of carbon dioxide below 1,000 ppm. The data indi-cate that the carbon dioxide levels are appropriate and that adequate amounts of outdoor air are supplied to the Tower.[39]

Formaldehyde. EH&E typically took samples from the nursing stations, which are located in the central area of each pod. The sampling was done predom-inantly during the day shift when the majority of the complaints were noted. The measured levels of formaldehyde were well below the most stringent exposure guidelines. The data collected were consistent with the measured levels of hydro-carbons, as the levels of formaldehyde in the spaces were slightly higher than the levels measured in the supply air and the outdoor air. This difference suggests that the quality of the outdoor air has less impact on the indoor formaldehyde levels than sources within the Tower, a typical pattern in buildings.[40]

Glutaraldehyde. EH&E again took samples from the nursing stations during the day shift. The measured levels of glutaraldehyde were below the detection limit of the method, which has an average limit of detection (depending on sam-ple air volume) 1,000 times lower than the occupational-exposure guidelines. No glutaraldehyde was detected, although an Exomat X-ray developing unit was lo-cated in the core of Floor 8, and that developer employs a fixative that contains glutaraldehyde.[41]

---

[37]Ibid. at 22; Environmental Health and Engineering, *Environmental Measurements Reference Sheet,* August 25, 1994, at 1.

[38]Environmental Health and Engineering, *Indoor Environmental Quality Investigation at the Patient Tower—Draft,* November 14, 1994, at 22–24.

[39]Ibid. at 24–26.

[40]Ibid. at 28–32.

[41]Ibid.

Total hydrocarbons. The total hydrocarbon measurement represents an integrated measurement of the concentration of organic compounds in the air sample. Specifically, the measurement reflects hydrocarbons such as benzene, toluene, and xylene. These hydrocarbons are measured to detect abnormally high concentrations that may prompt more detailed investigation and to detect the location of infiltration of hydrocarbon gases or vapors. EH&E continuously measured the total hydrocarbons during the period of January 1994 to April 1994. The minimum and average levels of hydrocarbons were consistent with those of other buildings of similar use. In addition, the levels measured in the supply air were generally lower than the levels measured within the workplace. EH&E explained the difference in these two measurements by the presence of cleaning and patient-care products used on each of the Tower floors. A few areas did have sporadically elevated levels of hydrocarbons. EH&E concluded that these peaks represented short-duration, isolated events that released a significant amount of material near the sampling tube (e.g., preparing an alcohol rub). Secondary measurements in the same areas reflected average values, supporting EH&E's conclusion.[42]

Bioaerosols. EH&E took biological air samples in March 1994 through May 1994 with an Anderson N6 impactor. EH&E detected low levels of bacteria and fungi, a typical finding in uncontaminated indoor environments. The bacteria and fungi detected were common species often found in indoor and outdoor environments and are not considered disease-causing. The levels of airborne fungi and bacteria in the Tower were substantially below any concentrations thought to be problematic.[43] In fact, the concentrations of fungi in the Tower were extremely low: the samples contained generally less than 13 cfu/m^3, and the highest measurement was 35 cfu/m^3. The samples taken at the HVAC air intakes averaged 195 cfu/m^3, indicating that the air filters were working effectively and there were no sites of amplification within the building.[44]

Volatile organic compounds. EH&E employed the use of SUMMA polished stainless-steel canisters to collect air samples. EH&E analyzed 42 target compounds through this SUMMA method and identified compounds that were typical for the normal hospital environment but below the most stringent health guidelines. EH&E identified four compounds that did exceed their respective odor thresholds: n-butane, ethanol, isopropanol, and d-limonene. Each of these is a common hospital product, and EH&E determined that none was the cause of an odor complaint.[45]

---

[42] Ibid. at 26–28.

[43] Ibid. at 33–35.

[44] Ibid. at 35.

[45] Ibid. at 38–40.

Latex. EH&E collected samples from settled dust to determine if latex allergen from surgical gloves used in the Patient Tower was building up. EH&E also collected two air samples. The settled-dust samples were collected where accumulations of dust were visible, as well as from the back of ceiling tiles. EH&E took samples from various floors, predominantly from Pod B and Pod C, where most of the complaints were occurring. The results from the two pods did not differ significantly. The samples were at or near the detection limit for the method used.

Medication Use

The environmental data indicated that local, as opposed to buildingwide, sources of airborne material emissions may be a factor affecting the air quality. One important element in this analysis was medication storage, use, and disposal. EH&E initially targeted antibiotics as a potential source of IAQ complaints. A potential source of the complaints was subsequently linked to the frequent use of ampicillin for newborns on the third and fourth floors of the Tower. Typically the staff mixes powdered ampicillin with a saline solution just before use, administers the proper dosage to the infant, and discards the remainder of the antibiotic. EH&E concluded that this technique was a possible source of some symptoms that employees on the third and fourth floors experienced. Medication use was also thought to correlate with other staff-reported symptoms on other floors of the Tower.[46]

EH&E identified three potential sources of complaints related to medication use: antibiotics, Albuterol (an aerosolized bronchodilator), and Recombivax HB hepatitis vaccine. Antibiotics, including Albuterol, have long been associated with hypersensitivity reaction in humans. Although there are few published studies concerning the effects of antibiotics on hospital staff, patient studies confirm that long-term use may result in hypersensitivity reactions that include rashes, dermatitis, throat irritation, eye irritation, headache, shortness of breath, and dizziness.[47]

In response to increased awareness of the potential effects of medication use, the Brigham adopted new procedures for the mixing and handling of antibiotics and other drugs. These procedures included disposing of unused medications and used syringes in covered wastebaskets and receptacles, avoiding the spraying of drugs from syringes to remove air bubbles, and priming intravenous (IV) drip lines by draining them into appropriate receptacles.[48]

---

[46] Ibid. at 11–12.

[47] Ibid. at 14–16.

[48] Ibid. at 15–16.

Corrective Measures

The Brigham took the following actions in response to the investigation of the Patient Tower:

• By May 1996, all of the floors of the Patient Tower were scheduled to be power cleaned. The Brigham cleaned the 3rd and 4th floors in the spring of 1994, the 12th floor in May 1995, and the 8th floor in August 1995. The remaining floors were cleaned through May 1996.

• The Brigham established an automated telephone survey system that permitted employees concerned about IEQ to call and record their symptoms at any time.

• The Brigham established a paging system so that one number (5000) would be used exclusively to handle responses to employees' reports of environmental quality incidents. An industrial hygienist carries the pager during business hours; a nurse administrator does so during off hours. The industrial hygienist or the nurse administrator fills out a form and notes the hospital's response to each reported incident.

• In March 1994, the Brigham instituted a standard policy governing the type of latex and nonlatex gloves that can be used and ordered through the purchasing system. This policy makes it impossible for departments to order nonapproved gloves.

• The Brigham hired an occupational health physician to oversee employee medical care and formed a department of occupational health. This department developed protocols for dealing with environmental cases and established a support group for employees affected by environmental exposures.

UNION ISSUES

Brigham and Women's Hospital's nurses are represented by the Massachusetts Nurses Association (MNA). Because the majority of employees reporting possible environmentally related symptoms were nurses, MNA assumed an advocacy role on behalf of nurses concerned about the indoor environment at the Brigham. In August 1993, after the cluster of symptoms was identified in the operating rooms, MNA requested and received from the Brigham information about the identities and work locations of individuals who believed that they had been affected, the work status of those employees, and the results of the environmental investigations for the work areas in which the symptoms were reported. These informational exchanges occurred through correspondence and regular meetings attended by Brigham and MNA representatives.

In addition, MNA requested, and the Brigham undertook, certain voluntary extracontractual commitments to alleviate MNA's concerns about the status of nurses who reported that their ability to work at the Brigham was adversely affected by their alleged symptoms. The Brigham extended certain time limits provided by the collective-bargaining agreement applicable to benefits concerning work-related leaves of absence as well as other benefits for those who reported that they could not work due to symptoms they believed were environmentally related. These extracontractual commitments were made as a result of a process separate from the contractual negotiations that started in September 1993, shortly after the OR investigation began, and concluded in January 1994.

As concerns about IEQ spread to other areas of the Brigham, MNA and Brigham representatives continued to meet to share information and to respond to each other's concerns. These meetings included sessions in which representatives of EH&E reported investigative findings and the Brigham's newly hired occupational health physician answered questions about whether particular symptom reporting was significant and possible means of conducting occupational health investigations.

In addition to meeting with MNA representatives about these issues, the Brigham and EH&E staff met regularly with employees, many of whom were nurses, in areas of the hospital where concerns had been raised or reports of symptoms had occurred. These meetings were scheduled so that employees from all shifts could attend.

Despite the fact that the environmental investigations indicated no identifiable cause of most of the reported symptoms, and despite the hospital's efforts to make improvements to an indoor environment that met or exceeded the standards set out by even the most stringent regulations and guidelines, MNA continued to believe that the indoor environment was unsafe. MNA went to the media in July 1994 with certain demands for indoor environmental improvements and monitoring, even though the Brigham had already agreed to many of these. In August 1994, the Brigham suggested that such an approach was counterproductive. The Brigham also expressed concern that much of the information it shared with MNA was preliminary and not ready to be publicized.

In December 1994, MNA requested that those of its members remaining on industrial accident leave as a result of alleged environmental exposures be permitted to supplement workers' compensation benefits with vacation, holiday, and sick time. Again, pending the outcome of additional environmental investigation and to demonstrate good faith, the Brigham agreed to this request. The Brigham also clarified the long-term disability benefits available to those who were receiving workers' compensation benefits on an accepted liability basis.

During the winter and spring of 1995, employees on the 12th floor of the Patient Tower began to report increasing numbers of symptoms. As a result, the hospital moved patients off the floor, conducted extensive environmental investigation, and made multiple improvements to the floor. The improvements included deep cleaning all areas, except above the ceiling tiles in three of four patient care pods. Above-ceiling cleaning was performed in Pod 12D where most of the symptoms had been reported. Within a few weeks of the floor's reopening, the nurses again began to report symptoms.

At about the same time, in April 1995, the Brigham and MNA asked a facilitator to work with them in an attempt to help make more productive their joint approach to concerns about IEQ. The facilitation process did not have any binding or precedential value but was instead an attempt to improve communication and achieve a more collaborative process. The facilitation proceeded in the form of frequent, lengthy meetings held over the next four months. During that time, the hospital again closed the 12th floor to reinvestigate possible causes of symptoms and to deep clean above the ceiling tiles in the three pods where such cleaning had not previously taken place. The Brigham and MNA also retained neutral outside experts to determine whether reentry to the 12th floor was safe. Following a review of the investigative data and reports, and meetings with MNA representatives, the Brigham concluded that reopening the floor was safe.[49]

Shortly after the successful reopening of the 12th floor, MNA demanded that the Brigham close the 8th floor due to increased symptom reporting. The hospital had already developed a plan for removing patients from this floor as part of its deep cleaning program and had, in response to the increased reporting, accelerated the schedule for such cleaning to take place. The accelerated schedule was developed after consideration of necessary lead time to appropriately equip the floors to which these particular patients would be moved. Nevertheless, MNA apparently went to the press with reports that the Brigham was not being responsive to its demands concerning the closure of the 8th floor. As a result, the Brigham decided that it would no longer participate in the facilitation process, although it continued to share with MNA, through correspondence, telephone calls, and meetings, information about environmental investigations and symptom reporting.[50]

During the next year, the number of symptom reports generally declined, and MNA and the hospital entered into contract negotiations, which concluded in September 1996. In May 1996, MNA led an informational picket demonstration outside the Brigham concerning IEQ. At the time that this chapter was submitted

---

[49]*See* Thomas J. Smith, Ph.D., and David H. Wegman, M.D., *Letter to John F. McCarthy,* June 12, 1995; Environmental Health and Engineering, Inc., *Report on Environmental Performance Date and Facility/Engineering Improvements Patient Tower—12th Floor Pods A, B, C, D, and Core Area* [Undated].

[50]Mary S. Fay, R.N., V.P., Nursing, *Letter to Staff Nurses Regarding Workplace Environmental Issues,* October 5, 1995.

for publication, the parties had concluded negotiations and had reached a contract agreement that included a provision that would provide limited retraining and compensation for certain eligible nurses who claim that their alleged environmental illness prevents them from working at the Brigham.

LEGAL ISSUES

It is important to note that the environmental issues at the Brigham have not affected any patients. The legal issues that have arisen from the workplace environment are purely employee-based.

Workers' Compensation Cases

From 1993 to 1996, the number of employees who reported complaints allegedly related to the IEQ at the hospital fluctuated, with peaks in reporting occurring in July 1993, February 1994, and August to December 1995. A review of the available data does not explain the rise and fall of the numbers of reports, but a few events are noteworthy. A comparison of the reported complaints and alleged lost time from work are set forth in Figures 15.4 and 15.5.[51] These graphs also identify some of the events that were occurring at the time of increased rates of reporting.

As of 1995, 100 employees had received or were receiving weekly workers' compensation benefits as a result of alleged environmental exposures at the Brigham that allegedly caused their claimed illnesses. The number of employees who began to receive ongoing weekly workers' compensation benefits dramatically increased during the period of November 1993 through April 1994. A review of the dates on which employees began to receive weekly workers' compensation is set forth in Figure 15.6.

Several of the cases shown in the graphs in Figure 15.6 are currently in litigation before the Massachusetts Department of Industrial Accidents (DIA), based on either the employee's claim for continuing weekly workers' compensa-

---

[51] The Brigham notes that the OSHA 200 logs have been completed by ruling as reportable substantially all incidents recorded on incident report forms and submitted to the Brigham's Safety Office. There has been virtually no attempt to distinguish incidents by the criteria elaborated in OSHA regulations, specifically 29 CFR § 1904.12. This means that the number of reported occupational injuries or illnesses may well be inflated.

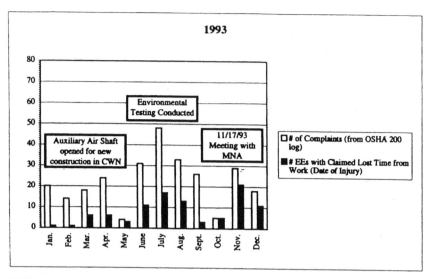

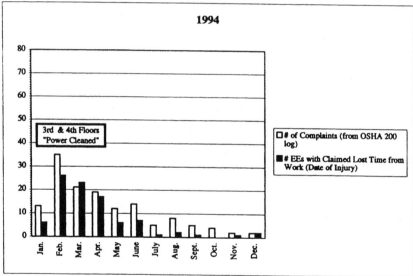

Figure 15.4. Worker's compensation statistics (based on OSHA 200 logs).

tion benefits or the Brigham's complaint to modify or discontinue the employee's benefits. Of the cases in litigation at the DIA, 26 are claims for benefits and 57 are complaints to modify or discontinue benefits. Of the total of 83 cases in litigation, 17 have settled and 8 have reached an interim status through an agreement to pay some type of weekly compensation benefits. In cases where

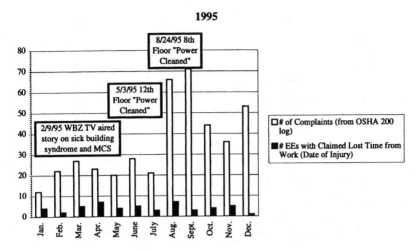

1995

Figure 15.5.   Workers compensation statistics (based on OSHA 200 logs).

DIA proceedings are scheduled or under way, 4 cases are at Conciliation,[52] 10 are at Conference,[53] and 31 cases are at the full evidentiary hearing stage involving sworn testimony and exhibits introduced in accordance with the Massachusetts evidence standards, as well as medical evidence through a report and deposition testimony by a DIA-selected impartial physician in an appropriate specialty.[54] Of the remaining cases, six are awaiting an impartial physician evaluation and report, and seven have been withdrawn by one of the parties.

Litigation results. Of the 26 claims for benefits the Brigham has obtained the following results at the Conference level: in 3 cases, DIA ordered medical benefits; in 2 cases, DIA awarded partial disability workers' compensation benefits; in 7 cases, DIA awarded temporary total disability workers' compensation benefits; and in 9 cases, DIA denied the employee's claim for benefits.

---

[52] The Conciliation is a procedure established by Mass. General Laws (M.G.L.) c. 152, § 10, in which a DIA Conciliator preliminary reviews the case and identifies issues for the parties. The Conciliator attempts to help the parties resolve the case if possible. The Conciliator will send the case forward to a Section 10A Conference if the moving party has produced adequate evidence in support of its claim or complaint.

[53] A Conference occurs pursuant to M.G.L. c. 152, § 10A. This process requires that an administrative judge evaluate the medical records, reports of injury, signed statements from the employee and any witnesses, rehabilitation records, and any other documents, as well as the oral arguments of the parties' attorneys. Following the Conference, the administrative judge must issue an order denying or allowing the employee's claim or the insurer's complaint.

[54] The DIA evidentiary hearing is conducted pursuant to M.G.L. c. 152, § 11.

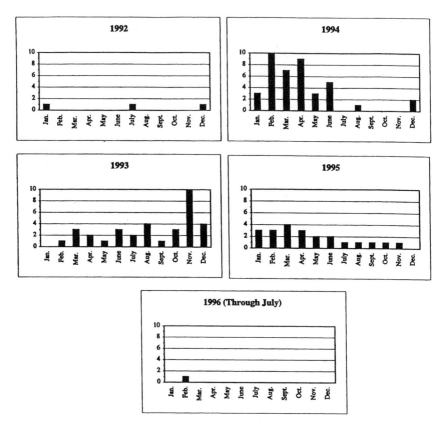

Figure 15.6.   Number of employees placed on ongoing worker's compensation.

Of the 57 complaints to modify or discontinue benefits, the Brigham obtained
the following results at the Conference level: in 15 cases, DIA ordered a modification (reduction) of the employee's weekly workers' compensation benefits; in 1
case, DIA ordered an outright discontinuance of the employee's benefits; and in
22 cases, DIA declined to reduce or terminate the employee's benefits. The results
that the hospital has obtained at DIA do not appear to correlate to any specific factors in the presentation of the case, including the submission of medical reports,
environmental sampling data and reports, or labor market survey analyses. Despite consistent case handling by the attorneys representing the Brigham at the
proceedings, DIA administrative judges have issued varying opinions. Factors
that do appear to affect the outcomes are surveillance reports that demonstrate that
the employee is engaging in activities inconsistent with his or her claim, and the
opinions of the impartial physicians on causation of the employee's symptoms
and the degree of the employee's disability, if any.

Multiple chemical sensitivities. The earlier claims resulting from alleged environmental exposures in the OR were based predominantly on alleged latex allergies, asthmatic reactions, bronchial hyperreactivity, and rashes that could be substantiated by objective medical testing. The employees' physicians could point to concrete data that substantiated the claim and documented a causal relationship between chemical agents present at the hospital and the employees' symptoms. As the Brigham experience progressed, more and more employees were complaining of symptoms that were not verifiable by the employees' physicians, and the hospital denied several of these unsubstantiated claims. Thereafter, several employees began to allege that they had developed MCS as a result of their environmental exposures at the hospital. MCS is a highly controversial diagnosis that has not gained general acceptance within the medical community (see Chapter 3 of this book). Several medical associations, including the American Medical Association, the American Academy of Allergy and Immunology, the American College of Physicians, and the American College of Occupational and Environmental Medicine, refuse to recognize MCS as a diagnostic disease.[55] The employees assert that, as a result of their exposures at the Brigham, they are unable to work in any environment that contains certain chemicals or odors. Affected employees claim that they can no longer go to grocery stores, malls, movie theaters, or other public areas because of an increased sensitization to perfumes, cleaning agents, tobacco smoke, diesel exhaust, and deodorizers. The majority of the employees asserting claims of MCS are being treated by holistic medicine physicians and clinical ecologists. These physicians are prescribing nonconventional treatments consisting of intravenous vitamin C, nutritional therapy, heat detoxification (sauna treatments), megadoses of vitamins, and restrictive diets. The hospital has consistently denied that these treatments are reasonable or medically necessary.

The Brigham also has challenged the diagnosis of MCS on the basis of the reasoning in *Commonwealth v. Lanigan,* 419 Mass. 15 (1994), in which the Supreme Judicial Court expressly adopted the federal evidentiary standard outlined in *Daubert v. Merrell Dow Pharmaceuticals, Inc.,* 509 U.S. 579 (1993), governing the admissibility of expert testimony. According to these cases, before a witness can testify competently as to his or her expert opinion on an issue, the court must make a threshold determination as to the scientific reliability of the opinion to be offered. Where the proponent of expert evidence fails to lay an adequate foundation for the scientific reliability of the opinion offered, then the

---

[55] See American Medical Association, Council of Scientific Affairs, Clinical ecology: Council report, 268 *JAMA* 3465, 3467 and nn. 1–3 (1992); ACOEM statement on multiple chemical hypersensitivity syndrome, multiple chemical sensitivities, environmental tobacco smoke, and indoor air quality, *ACOEM Report* (Feb. 1993); Clinical ecology, 78 *J Allergy Clin Immunol* 269, 269–71 (1986), reprinted C. Banov, L. Mendelson, J. Seiner & A.I. Terr, Position statement: Management of the patient with environmental illness, 4 *News and Notes* 8, 15 (1990).

opinion does not qualify as expert evidence and may not be considered to prove the scientific or medical issue in dispute.

In determining whether an expert's opinion has a sufficiently reliable basis to be admissible under *Lanigan,* the court must examine the scientific validity of the reasoning or methodology underlying the testimony.[56] Pertinent considerations in making an assessment under *Lanigan* include (1) whether a theory or technique can be and has been tested; (2) whether the theory has been subjected to peer review and publication; (3) the error rate of the technique; (4) whether there exist standards that control a technique's operation; and (5) whether the technique has gained general acceptance within the expert's field.[57] The Brigham has challenged the diagnosis of MCS on the bases that MCS has not gained general acceptance within the medical community, that MCS has not been subjected to peer review and publication, and that the MCS diagnosis has not been supported by any double-blind, controlled, peer-reviewed studies. The cases where the *Lanigan* challenge to the MCS diagnosis was used have not yet been decided by the DIA administrative judges who heard them.

• *Third-party litigation.* Several employees are pursuing actions against the building and ventilation contractors and subcontractors. The cases are currently pending in the Suffolk Division of the Superior Court Department of the Trial Court. Discovery in these cases is proceeding at this time.

• *Direct suit by 15 employees against the Brigham.* In July 1996, 15 Brigham employees filed a complaint directly against the Brigham and Women's Hospital, the Brigham Medical Center, Inc., and Partners HealthCare System, Inc. The employees make claims of strict liability, gross negligence, and negligence against all three defendants. The Brigham intends to test these claims in a motion to dismiss based on M.G.L. c. 152, §§ 23 and 24, which bar an employee from bringing a direct action against an employer or its insurer for injuries suffered in the workplace if the employee received or was eligible to receive workers' compensation benefits.

Practical Recommendations

Maintain lines of communication. A review of the informal complaints to OSHA and follow-up letters reveals that some staff at the Brigham perceived a commu-

---

[56]*Lanigan,* 419 Mass. at 26.

[57]*Lanigan,* 419 Mass. at 25–26; *Daubert,* 509 U.S. at 592–95.

nication problem between management and staff. For example, although the Brigham's industrial hygienists met regularly with the nursing staff, MNA representatives repeatedly reported to the Brigham that its members felt uninformed. This problem resulted in a sense of mistrust on the part of the staff that only escalated concerns. There also appeared to be some apprehension regarding potential employee layoffs due to the merger of the hospital to form Partners Community Health Care, Inc. This apprehension would of course be a significant source of stress that could contribute to certain employees' health problems. Communication should be complete and accurate to avoid escalation of problems associated with IEQ complaints.

Legal Practice Tips

With respect to handling the legal aspects of an IAQ problem, the following suggestions are offered:

• Immediately following a reported exposure by an employee, obtain a medical and hospital records release from the employee. Request copies of the employee's medical records to determine if the employee's complaints are new or a result of a preexisting condition or some other cause.

• Obtain a chronology of the employee's work history. This can come from the employee's personnel record and the employee's supervisors. The history could provide useful information concerning employment problems that provide an incentive for the employee to make a claim, or problems that the employee has alleged at other workplaces.

• Immediately conduct an investigation of the alleged incident or exposure after the employee reports it. Not only will this indicate to other employees that the employer takes all complaints seriously, but it results in the collection of valuable data in the event the claim is ever presented to a tribunal.

• Obtain past records of all environmental testing that has been done in the areas where the employee has been working, and schedule additional testing as soon as possible.

• Determine whether/additional complaints have come from the employee's work area. Cases from similar areas should be treated similarly. The Brigham soon discovered that information was being channeled to various employees regarding the testing that had been completed and the actions that were being taken with respect to other employees' claims. Employer consistency in managing air-quality complaints can be a significant aid to the outcomes of workers' compensation litigation.

• Send a complete set of the employee's medical records and test results to an independent medical examiner (IME) for review and comment. Determine what portion of the medical treatment, diagnosis, and prognosis is legitimate or generally accepted within the medical community. Investigate the laboratories where the employee's blood testing is being conducted to determine if they are properly licensed or certified and whether the employee's treating physician has any financial connections to the laboratory.

• Conduct investigation or surveillance when appropriate. This sort of inquiry is particularly helpful to determine whether the employee is engaging in activities that are inconsistent with the employee's claim; e.g. sampling and purchasing perfume at the department store, making home renovations, attending indoor sporting events, or going to the hair dresser.

CONCLUSION

This study of the Brigham experience is offered to show the complex interactions that occur when an indoor environmental health issue arises. Each situation is unique, but the levels of management attention, consultant expertise, and legal support that this case demonstrates can illustrate for building owners and managers the degree of effort that can be involved in resolving employee complaints.

Chapter 16

Case Study: Anchorage, Alaska

ALAN HEDGE, PH.D.

This case study describes an indoor air quality (IAQ) incident in Anchorage, Alaska, that began in 1985 and ended in 1986 with evacuation and litigation (Hedge et al., 1987).

An air-conditioned office building with operable windows was leased to a state agency and housed over 450 state employees. In 1985, following a water leak from a water heater, two female workers complained of respiratory problems. Both workers saw a physician, who diagnosed a possible mold allergy and suggested to them that they may be allergic to something in their building.

In early 1986, the building was investigated by state industrial hygienists and by consultants from private companies. The focus of the investigation was possible microbial contamination of the building. Nothing exceptional was found by any investigator. The search for a microbial cause continued. It was unfortunate that a misunderstanding arose when, during their visit to the building, the industrial hygienists wore full protective suits in anticipation of possible asbestos exposure. The sight of protective suits being worn in their building caused alarm among employees. Most workers had been unaffected until that time, but curious workers called news media about the investigation, and rumors began to circulate that the building suffered from a "mystery bug." The media began detailed coverage about the building problems.

Two months after the start of this incident in 1985, another state agency decided to conduct a questionnaire survey of all workers in the building as well as workers in a comparable building that had not reported any problems. The extensive questionnaire contained a number of leading questions. Over a one-week period, each worker in each of the two buildings received the questionnaire.

During the time that the questionnaire was being circulated, local news media ran scare stories about the building. Many workers in the problem building began to express concerns and increased anxiety about their safety at work. By this time, the investigation was out of control, and management had lost the trust of the employees.

While the survey results were being processed, one affected worker who complained of fatigue decided to lie down on the floor of a storage room. A relatively short time later, this worker was gasping for breath. An ambulance was called. A state epidemiologist was notified that the person had a severe adverse respiratory reaction to the building. The epidemiologist was already aware of preliminary results from the questionnaire survey of the problem building; these results showed a fairly high prevalence of symptoms of sick building syndrome (SBS). The epidemiologist reacted swiftly by having the building evacuated, and announced to the media that the evacuation was a precaution while investigators continued to search for the mystery bug.

Three crucial pieces of information were *not* available when the decision was made to evacuate the building:

1. Results of an evaluation of ventilation-system performance
2. Results of a comprehensive assessment of possible microbial contamination
3. Results from the questionnaire surveys that compared the two buildings using proper statistical methods

After the evacuation, experts were brought in to review previous findings, assess ventilation-system performance, conduct further tests for microbial contamination, and reanalyze the questionnaire survey data.

The outcome of subsequent work showed that, other than minor problems with the ventilation-system performance that could have affected the employees' thermal comfort, no evidence existed that the ventilation system was contaminated or was the source of the problem.

No evidence was ever found of any significant microbiological contamination. Reanalysis of the survey data showed that errors had been made that inflated the apparent problems in the building. When proper statistical tests were used, they found no significant differences between workers in the problem building and those in the control building. In later litigation, the correctly analyzed results from the questionnaire survey played a prominent role.

The building was reoccupied about six months later. Several lawsuits were filed, and the ensuing litigation spanned five years. None of the worker cases ever made it to trial. The building owner was ultimately vindicated but went bankrupt.

The sequence of events described in this case illustrates the dangers of responding to an IAQ issue by implementing an unplanned and poorly managed protocol such as a survey. In this case, events transpired that quickly led to a negative perception about IAQ. Everyone involved in managing the investigation accepted the first complaints at face value. All were convinced that a mystery bug was responsible and would be found. The manner in which subsequent IAQ investigations were conducted exacerbated the fear of potential hazards within the building. Management quickly lost control of what, when, why, and how investigations would be carried out.

As with many other SBS cases, local news media used a "shock horror" coverage that fanned the flames of concern. Managers were unskilled in dealing with the media. The investigators readily speculated about many hidden dangers that might be responsible for the problems. Decisions were made based on speculation rather than on proper collection of data.

Analysis of the chronology of complaints showed a distinct pattern to the transmission of anxiety and of SBS symptoms from worker to worker. In their rush to find the mystery bug and successfully solve this case, state investigators made simple errors in data analysis that overinflated the apparent sickness of the building.

Group contagion about a mystery bug affected workers and investigators alike. Everyone involved in the early stages of the case took what each considered to be the right actions, for the best of reasons, but they neglected to question what the real problem might have been, or whether any problem existed at all.

The lesson learned from this case study is that any SBS/IAQ investigation must be a carefully planned and managed activity, regardless of pressures to take action immediately and solve the problem quickly. The investigation needs to follow a systematic investigative methodology, the leader must coordinate the work of experts and investigators from a variety of disciplines, the leader must successfully communicate with and gain the trust of workers, and the management group must skillfully handle communications with the outside media. When approached in this way, any SBS/IAQ incident can be resolved to the satisfaction of all, without the potentially destructive aspects of a legal response to the SBS allegations.

REFERENCES

Hedge, A., et al. (1987) Indoor air quality as a psychological stressor. In B. Siefert et al, eds., *Indoor Air '87: Proceedings of the 4th International Conference on Indoor Air Quality and Climate, Berlin (West), 17–21 August,* Vol. 2, pp. 552–556.

Chapter 17

The Chicago High-Rise IEQ Case

ROBERT L. GRAHAM AND CYNTHIA A. DREW [1]

INTRODUCTION

The John Hancock Center in Chicago towers above its neighbors in the elite shopping district of North Michigan Avenue. At 100 stories and 1,127 feet high, it is considered one of the preeminent symbols of the Chicago skyline.[2] However, most people who today bustle by the John Hancock Center on the sidewalk or stop to gaze up its great heights are unaware that, some 10 years ago, this building was the focus of an indoor environmental quality (IEQ) lawsuit brought by the city of Chicago against the Center's owners, property manager, and condominium association.

A review of the facts of the John Hancock case, and of the attempts of certain John Hancock Center residents to obtain remedies for IEQ problems, offers instructive lessons for today's building managers, especially those who seek to address IEQ problems effectively before the mature into costly lawsuits. For a

[1] Robert L. Graham is chair of Jenner & Block's environmental law department, Chicago, Illinois. Cynthia A. Drew is an associate in Jenner & Block's environmental law department.

Robert Graham represented Ms. Joan Whitmer, a condominium owner in Chicago's John Hancock Center, in her successful indoor environmental quality suit to intervene in a case the City of Chicago brought against the owners, the property manager, and the condominium association of the John Hancock Center. The lawsuit alleged that the Center violated the city's building code.

[2] The John Hancock Center is the third-tallest building in Chicago, after the Sears Tower and the Amoco Building. Its location on North Michigan Avenue and its unique sloping shape make it a particularly distinctive example of Chicago's outstanding architecture.

time, the John Hancock Center controversy threatened to taint the reputation of the building the popular press dubbed "the ebony smokestack."[3]

FACTUAL BACKGROUND

Like many other high-rises in Chicago, the John Hancock Center combines commercial and residential uses in one building and contains a parking garage on the lower levels. The Center's offices and residences essentially occupy two different structures. The residential structure is built atop the commercial one.

In the context of launching a $2 million renovation program designed to save energy costs, Charles F. Clarke Jr., executive vice president of Sudler & Co., the John Hancock Center's property manager, explained, "In tall buildings like ours, there's a tremendous stack effect that pulls in frigid outside air every time the garage door opens.[4] This stack effect—the phenomenon of air being drawn from the lower commercial levels of the John Hancock Center to the upper residential levels—sometimes manifested itself dramatically. "Blasts of air from the elevator shafts have been known to knock off men's hairpieces and blow women's skirts up around their ears," Clark said. "It also moves upward through stairwells, electrical conduits, space around the plumbing, and gaps in the poured-concrete floors around the support beams."[5] This stack effect also transmitted upward carbon monoxide and other fumes originating in the parking garage as well as potentially dangerous air emissions created by the use of certain chemicals in the office units on the building's lower level. The effect was exacerbated by the speed at which the elevators shoot skyward to the observation deck and restaurant at the very top of the John Hancock Center—a rate of more than 1,000 feet per minute—and the buildings' lack of sufficient natural ventilation. By 1984, air quality at the John Hancock Center had for some time been attracting the concerned attention of the city of Chicago and many building residents.

According to applicable City of Chicago building code provisions, it was permissible for the office building portion of the John Hancock Center to have mechanical ventilation and windows that do not open. However, the residential portion of the Center was not permitted to have solely mechanical ventilation. Building code provisions required that a residential building have the equivalent of at least 5 percent of its floor space as windows that open to provide natural ventilation. The city's code further specified that these windows must be held in place by hardware and must be easy to open.

---

[3] Ver Berkmoes, "The Ebony Smokestack," *Chicago Lawyer,* Vol. 7, No. 6 (June 1984).

[4] Ibata, "Hancock Aims at $12 Million Energy Savings," *Chicago Tribune* (February 17, 1985).

[5] Ver Berkmoes at 5.

The residential apartments in the John Hancock Center had at least 5 percent of their floor space as windows. The windows and hardware shown on the plans the building's owners provided to the city in July 1965 would, in the city's view, have qualified as windows that were held in place by hardware and were easy to open. According to the city, however, the John Hancock Center did not install the windows or the hardware shown on the 1965 plans that the city approved. Instead, different windows and hardware were actually installed, and the city later alleged that these did not qualify under the code as windows held in place by easy-to-open hardware. Only the "hoppers" installed underneath every other larger window in the residential portion of the building could provide any natural ventilation to most apartments. However, the hoppers opened less than half an inch, providing insufficient natural air.

In addition, large air ducts from the building's mechanical ventilation system, also shown on the approved plans, were omitted during construction of the residential apartments as a cost-cutting measure. Instead, one to two inches were sawed off the bottom of each apartment door to create a pathway by which mechanical ventilation from ducts built in the building's corridors could reach the apartments. As a result, the only other fresh air most apartments received came from under their doors. Although relevant portions of the city's building code required such mechanical ventilation systems to be operated 24 hours a day, at the time of the city's lawsuit, the building had allegedly been operating its mechanical system as little as 13 $\frac{1}{2}$ hours a day.[6]

THE CITY'S LAWSUIT

Learning of the discrepancies between the original plans and the actual construction, the City of Chicago sued the John Hancock Center's owners, property manager, and the condominium association. The lawsuit alleged that the John Hancock Center had "failed to keep the mechanical ventilating system in good repair so as to ensure the required ventilation during all hours of occupancy" in specified areas of the building, had failed to "provide ventilation to ensure minimum air quantities as required by the Chicago Building Code" in "habitable rooms in all dwelling units," had failed to provide "ventilating openings equal in area to not less than 5% natural ventilation per square foot of floor area required in all dwelling units," and had failed to provide "a proper corridor air supply system to ensure that all kitchen and toilet exhaust fans can operate properly during all hours of occupancy in all dwelling units." The city asked for both a temporary and a permanent injunction requiring the defendants to correct the alleged viola-

---

[6] Ver Berkmoes at 1.

tions and to restrain future violations; the appointment of a receiver to correct the conditions alleged in the complaint "with the full powers of receivership, including the right to issue and sell receiver's certificates"; and a judgment against the defendants, including an order for the plaintiff "to demolish said premises under the police power to protect the public health and safety" and a lien on the property for the cost of demolition.

After a trial on certain of the disputed issues, an Illinois trial court judge issued two orders in the case. In the first order, partial judgment was entered in favor of the defendant condominium association, stating that the evidence heard at trial demonstrated "that the windows in the residential portion of the John Hancock Center are openable and satisfy the 5% natural ventilation requirement" of the relevant provisions of the City of Chicago's code.[7] In the same order, partial judgment was then entered in favor of the plaintiff City of Chicago "in that the windows do not presently have hardware which would allow the windows to be opened as required" by the city's building code. Consequently, the judge ordered the defendant condominium association to submit to the court and to the city within 30 days "a sample of the window hardware and installation specifications" that the defendant homeowners' association "proposes to be installed to cure the lack of compliance." In the second order, the judge dismissed the case after the court determined that it appeared the "ventilation system is not being operated" and the defendants "agreed to continue to operate the mechanical ventilation system" in accordance with the city's code.[8]

JOAN WHITMER'S INTERVENTION SUIT

While the city's lawsuit was still pending, Joan Whitmer, a John Hancock Center resident, sought to intevene, that is, to become an active participant, in the city's suit. At the time of Ms. Whitmer's attempted intervention, her husband Frank had already filed his own personal injury lawsuit (discussed later in this chapter). The Illinois trial court judge initially ruled that Ms. Whitmer could not intervene in the city's suit because her interests were supposedly adequately represented by the city and the condominium association. Setting a valuable precedent for the rights of condominium owners, the Illinois appellate court reversed this ruling following an expedited appeal.[9]

---

[7] May 29, 1984 Order.

[8] May 29, 1984 Order.

[9] *City of Chicago v. John Hancock Mutual Life Insurance Company,* 127 Ill. App. 3d 140, 468 N.E. 2d 428 (Ill. App. 1 Dist. 1984).

The appellate court determined that Ms. Whitmer should be allowed to intervene because neither the city nor the homeowners' association could adequately represent her interests. According to the apellate court, the city's representation in the existing lawsuit was "shaped by competing pressures from a large and diverse constituency" that may not represent Ms. Whitmer's interests.[10] Similarly, the court said the homeowners' association may not represent the views of all of its members. As the appellate court noted, the Center's condominium association "denies the existence of a building code violation, and in what can only be seen as a study in understatement, characterizes its position relative to that of [Ms. Whitmer] as 'a difference of opinion.' "[11] The appellate court analogized Ms. Whitmer's requested intervention to that allowed in a zoning case, where adjacent landowners are allowed to intervene when their particular interests extend beyond the public interest.[12]

At the time Ms. Whitmer's case was sent back to the trial court judge, the city was still taking discovery in its lawsuit against the John Hancock Center. The appellate court's ruling allowed Ms. Whitmer to participate in all further proceedings, including the subsequent trial, as an equal party. Essentially, the ruling in Ms. Whitmer's favor ensured that the case would go to trial. From that point on, even if the city agreed to settle its suit, the trial court judge could not approve the settlement unless the unhappy resident also settled. She did not.[13] As noted previously, after a trial of certain disputed issues, the court ordered the John Hancock Center to replace the hardware on the windows in all its residential apartments.

FRANK WHITMER'S PERSONAL INJURY LAWSUIT

Even before the city sued the John Hancock Center, Frank Whitmer, Joan Whitmer's husband and the former chair of the condominium's engineering committee, had brought a personal injury action against the Center. Mr. Whitmer alleged that carbon monoxide and other pollutants to which he was exposed during the time the couple lived in their John Hancock Center apartment had caused his res-

---

[10] *Chicago v. Hancock,* 468 N.E.2d at 433.

[11] *Chicago v. Hancock,* 468 N.E.2d at 433

[12] *Chicago v. Hancock,* 468 N.E.2d at 432.

[13] The type of settlement demands the parties considered included the acquisition of the Whitmers' condominium unit, the provision of major medical health insurance coverage to the Whitmers, payment of Ms. Whitmer's attorneys' fees and costs; and her damages and other related costs.

piratory and heart conditions and nerve damage.[14] The settlement issues addressed in his case included medical expenses, lost business or investment income or capital, maintenance costs for the Whitmers' condominium, out-of-pocket expenses for business upkeep, and lost pension benefits due to early retirement.

The ability of dissatisfied residents and unresolved lawsuits to create public-relations headaches for building managers is illustrated by a news release disseminated by architect Kenneth R. Woods in cooperation with Mr. Whitmer and Dr. Theron G. Randolph. At an October 1984 meeting of the American Academy of Environmental Medicine (AAEM) in Chicago, Woods cited "document after document" allegedly showing "massive violations of the Chicago Building Code that began while the John Hancock Center was under construction in 1969 and 1970." In his October 15, 1984, news release, Woods reiterated that "These violations have not yet been corrected some fifteen years later, even after a successful suit to enforce the Code by the City of Chicago against owner-developer John Hancock Mutual Life Insurance Company, managing agent Sudler and Company, and the condominium association."[15]

Woods attached to his news release three pages of excerpts from presentations made to AAEM entitled "Sick Buildings Make Sick People." According to Woods,

> A classic example of a "sick building" is the John Hancock Center in Chicago. An all-electric world-recognized building with a seven-story enclosed parking garage, almost fifty floors of offices, commercial and service floors, with another fifty-some apartment floors above. A building whose residential windows are without hardware to make them usable, and whose limited residential mechanical ventilation systems were then also turned off almost eleven hours a day to save money.

In addition, Woods also disseminated with his news release a two-page summary entitled "Violations of the Chicago Building Code Requirements for Natural and Mechanical Ventilation by the John Hancock Center," which he claimed listed the history since 1969 of the city's citations of the Center for building code violations.

POSTSCRIPT

Although the John Hancock Center made various offers to settle the Whitmers' lawsuits, no settlement was reached. Ultimately, to eliminate the stack effect, the

---

[14] In 1981, the Whitmers had moved from their John Hancock Center apartment to a home in Naperville, Illinois.

[15] Appendix A, October 15, 1984 Woods and Associates news release.

John Hancock Center completely reconstructed the ventilation system in its enclosed parking garage to vent it to the outside. As noted previously, in accordance with the court order, the Center also eventually changed the hardware on the windows in all its residential apartments. It is estimated that these improvements cost hundreds of thousands of dollars.[16]

AVOIDING LAWSUITS IN IEQ SITUATIONS: LESSONS LEARNED

As noted previously, people who seek to address IEQ problems effectively before these issues mature into costly lawsuits may profit from the experiences of those who found themselves enmeshed in the John Hancock Center's IEQ controversy. Following are some key lessons that building managers can take away from this account of the ebony smokestack.

1. *An ounce of prevention is worth a pound of cure.* Other portions of this book have recommended that even a single IEQ complaint be taken seriously, investigated thoroughly, and, if at all possible, resolved before it engenders more IEQ complaints. In the John Hancock Center case, the city had cited the Center for building code violations and residents had complained to the Center about IEQ problems for many years before the city eventually filed suit.

2. *To avoid potential IEQ problems, be vigilant in ensuring that construction proceeds according to the original specifications.* As a corollary principle, when IEQ issues arise, check the original specifications to determine if any changes in construction may have created IEQ problems. In the John Hancock Center case, significant deviations from the original specifications occurred during construction, and many of these increased the likelihood that the building would subsequently develop IEQ problems. For example, decreasing the amount of natural ventilation to residential apartments exacerbated the IEQ issues likely to arise in a building with a pronounced stack effect and an enclosed indoor parking garage.

3. *When IEQ problems arise, don't ignore them.* Other portions of this book have also emphasized the importance of addressing head-on any IEQ problems that arise, preferably with a competent team of interdisciplinary professionals. As the John Hancock Center case illustrates, a penny saved is not always a penny earned. The architect's project man-

---

[16] On cross-examination at trial, the project manager for the John Hancock Center's architect, Skidmore Owings & Merrill, testified that the estimated cost in 1970 of the hardware orginally planned for the windows had been $50,000.

ager testified at trial that the hardware eventually installed on the Center's windows in 1985 could have been installed in 1970 for approximately $50,000. However, in the mid-1980s, after the Center had already incurred significant legal fees in defending the city's suite, the Center then had to bear the substantial expense of retrofitting the building to address IEQ issues.

4. *The interests of a building association may be different from the interests of residents and users with IEQ problems.* A building association may be reluctant to authorize a building manager to use funds in retrofitting units to address IEQ problems. However, in such matters, building managers may find it impossible to satisfy all the interests of divergent stakeholders. In the John Hancock Center case, the Illinois appellate court found that the potentially competing interests of a resident with IEQ problems and those of the condominium association justified allowing an individual right of intervention in an IEQ case.

5. *Individual unit owners have the right to file lawsuits concerning impacts to their own health caused by IEQ problems and may do so if they do not believe that building owners or managers are addressing their IEQ concerns.* Although the condominium association may pay a building manager's salary, an association that understates a resident's IEQ concerns as merely a difference of opinion may find itself attempting to defend that position in court. In the John Hancock Center case, the condominium association lost that battle. A wise building manager will seek to resolve such differences and provide concerned residents with IEQ solutions that forestall costly litigation.

6. *The equitable remedies available to either government entities or private individuals in IEQ suits cover a broad range.* A governmental entity has an extremely broad range of equitable remedies it can pursue in court against an offending building for alleged code violations. In the John Hancock Center case, the equitable remedies the city originally requested included an order for the plaintiff "to demolish said premises under the police power to protect the public health and safety" and a lien on the property for the cost of demolition. It is extremely unlikely that a judge would go as far as granting an order of demolition, even to a governmental plaintiff. However, in the John Hancock case, the judge did order the defendant homeowners' association to retrofit the building as the city requested. The homeowners' association then had to bear the greatly increased costs, through assessment, of a construction decision made some 15 years earlier to change specifications.

Similarly, a private individual whose health, safety, or welfare is allegedly being diminished by the operation of a building may also go to court seeking a wide range of remedies, including money damages. Potential remedies for an intervenor may include the sale of a condominium unit, provision of major medical health insurance coverage, payment of attorneys' fees and costs, and other money damages and related costs. Potential remedies for a personal-injury plaintiff may include medical expenses, lost business or investment income or capital, maintenance costs for a condominium unit, out-of-pocket expenses for business upkeep, and lost pension benefits due to early retirement.

CONCLUSION

At one time in the mid-80s, no less than three lawsuits were pending simultaneously against the "ebony smokestack" to remedy the effects of alleged code violations. The City of Chicago's original lawsuit against the John Hancock Center resulted in a court-ordered retrofitting of the hardware on the windows of all residential apartments. The individual intervenor's lawsuit established an important precedent in Illinois courts that homeowners' associations do not necessarily represent the interests of individual unit owners when health and safety violations are alleged and that individual unit owners are therefore independently entitled to their own day in court. The broad-ranging nature of the individual remedies available to personal-injury plaintiffs underscores for today's building managers the importance of seeking an early and satisfactory resolution of all IEQ complaints. Building managers who can profit from the IEQ lessons outlined in this chapter should be able to avoid the legal and public relations problems that at one time threatened to taint one of Chicago's most prominent architectural landmarks.

Additional Reference Materials
on Indoor Environmental Health

As expressed earlier in the text, seek out competent and experienced legal counsel for advice on specific rights and liabilities. These references are not intended as legal advice or as endorsements of the conclusions reached by authors of these legal materials.

U.S. Consumer Product Safety Commission, *An Update on Formaldehyde* (October 1990)

U.S. Consumer Product Safety Commission, *Biological Pollutants in Your Home* (1/90)

U.S. Dept. of Justice, Americans with Disabilities Act, 28 Code of Federal Regulations, Part 35

U.S. Environmental Protection Agency (EPA), *Introduction to Indoor Air Quality: A Reference Manual*, EPA/400/3–91/003 (July 1991)

U.S. EPA, *Introduction to Indoor Air Quality: A Self-Paced Learning Module*, EPA/400/3–91/002 (July 1991)

U.S. EPA and American Lung Association, *Asbestos in Your Home* (booklet), (9/90)

U.S. EPA and American Lung Association, *What You Should Know About Combustion Appliances and Indoor Air Pollution* (9/90)

U.S. EPA, *Second Hand Smoke*, EPA 402-F-93–004 (July 1993)

U.S. EPA, *Targeting Indoor Air Pollution: EPA's Approach and Progress*, EPA 400-R-92–012 (March 1993)

U.S. EPA, *The Inside Story: A Guide to Indoor Air Quality*, EPA 402-K-93–007 (April 1995)

U.S. EPA and American Lung Association, *Indoor Air Pollution: An Introduction for Health Professionals*, GPO 1994–523–217/81322

U.S. EPA, *Current Federal Indoor Air Quality Activities,* EPA 402K-95005 (June 1995)

U.S. EPA, *Carpet and Indoor Air Quality Fact Sheet* (10/92)

U.S. EPA, *Indoor Air Quality in Public Buildings,* EPA 600/S6–88/009a (Sept. 1988)

U.S. EPA, *Setting the Record Straight: Secondhand Smoke Is a Preventable Health Risk,* EPA 402-F-94–005 (June 1994)

U.S. EPA, *Indoor Air Facts No. 4 (Revised), Sick Building Syndrome* (ANR-445-W, April 1991)

U.S. EPA, *Ventilation and Air Quality in Offices Fact Sheet,* EPA 402-F-94–003 (July 1990)

U.S. Department of Labor, Occupational Safety and Health Administration, *Proposed Rules on Indoor Air,* 59 Fed. Reg. 15968 (Apr. 5, 1994)

U.S. Department of Labor, Occupational Safety and Health Administration, *Final Rules, Asbestos and Building Owner Duties,* 59 Fed. Reg. 40694 (Aug. 10, 1994)

RELATED COURT DECISIONS

Appellant v. Respondent, 1993 Westlaw 17189 (Tex. Work. Comp. Com.) [MCS proof standard for compensation]

Bellsouth Telecommunications Inc. v. W.R. Grace & Co., 77 F.3d 603 (2d Cir. 1996) [time action accrued for costs of building product removal from headquarters building]

Chanin v. Eastern Virginia Medical School, 20 Va. App. 587, 459 S.E.2d 523 (1995) [MCS not established as arising out of claimant's employment]

Conradt v. Mount Carmel School, 539 N.W.2d 713 (Wisc. App. 1995) [MCS claims denied]

Dobbs v. Board of Supervisors of Louisiana State University, 588 So.2d 764 (La. App. 1991)

Flue & Tobacco Cooperative Stabilization Corp. v. EPA, No. 6:93CV00370 (M.D.N.C. filed 1994), discussed in 24 Occupational Safety & Health Reports (BNA) 486 (Aug. 3, 1994) [challenge to EPA classification of secondhand smoke as carcinogen]

Gupton v. Commonwealth of Virginia, 14 F.3d 203 (4th Cir., 1994) [no due process right to sue employer for smoke-free environment]

Hennly v. Richardson, 264 Ga. 355 (1994); *Pechan v. DynaPro Inc.* 622 N.E.2d 108 (Ill. App. 1993) [workers' compensation bars suit for exposure to secondhand smoke]

Johannsen v. New York City Department of Housing Preservation and Development, 84 N.Y.2d 129, 638 N.E.2d 981 (1994) [workers' compensation coverage of injury from secondhand smoke]

Kensell v. Oklahoma, 716 F.2d 1350 (10th Cir., 1983) [no "right" to smoke-free environment]

McCreary v. Industrial Commission, 172 Ariz. 137, 835 P.2d 469 (Ariz. App. 1992) [MCS not occupational disease for computer engineer]

Palmer v. Del Webb's High Sierra, 838 P.2d 435 (Nev. 1992) [workers' compensation denial for exposure to secondhand smoke in casino workplace]

Roberts v. Estate of Barbagallo, 531 A.2d 1125 (Pa. Super. 1987) [home purchaser complaint for undisclosed insulation gas; rescission granted]

Rutigliano v. Valley Business Forms, 1996 Westlaw 375429 (D NJ 1996)

Schober v. Mountain Bell Telephone Company, 600 P.2d 283 (N.M. Ct. App. 1978) ["accident" included exposure to smokers]

Shimp v. New Jersey Bell Telephone Company, 368 A.2d 408 (N.J. Super. 1976) [injunctions granted for secondhand smoke abatement after suit by employee]; see contra, *Gordon v. Raven Systems and Research Inc.,* 462 A.2d 10 (D.C. App. 1983)

Smith v. Western Electric Company, 643 S.W.2d 10 (Mo. App., 1982); State ex rel. *A&D Limited Partnership v. Keefe,* 77 Ohio St. 3d 50 (1996) [SBS claim, denied by jury, reinstated by judge]

PERIODICAL AND BOOK REFERENCES

American Bar Association, *Indoor Pollution and Sick Building Claims—Are They Covered Despite Absolute Pollution and Contamination Exclusions?* Tort and Insurance Practice Section, Annual Meeting Program, August 1996.

Frank B. Cross, *Legal Responses to Indoor Air Pollution* (Quorum Books, 1990)

James R. Davis and Ross Brownson, *A Policy for Clean Indoor Air in Missouri: History and Lessons Learned,* 13 St. Louis U. Pub. L. Rev. 749 (1994)

Lawrence Ebner, Defending the Industry Against MCS, *Pest Management* 14 (Oct. 1992).

Jean Eggen, Toxic Reproductive and Genetic Hazards in the Workplace: Challenging the Myths of the Tort and Workers' Compensation Systems, 60 Fordham L. Rev. 843 (1992).

Mark Gottlieb, Second-Hand Smoke and the ADA: Ensuring Access for Persons with Breathing and Heart Disorders, 13 St. Louis U. Pub. L. Rev. 635 (1994).

Grace Giuffrida, The Proposed Indoor Air Quality Acts of 1993: The Comprehensive Solution to a Far-Reaching Problem? 11 Pace Envtl. L. Rev. 311 (1993).

Andrew Harrison Jr., An Analysis of the Health Effects, Economic Consequences and Legal Implications of Human Exposure to Indoor Air Pollutants, 37 S.D. L. Rev. 289 (1992).

Catharine Hanrahan and John Beiers, Indoor Air Quality in the Office Environment: An Expanding Area of Toxic Tort Litigation, Toxics Law Rep. (BNA) 142 (July 9, 1994).

Gene J. Heady, Stuck Inside These Four Walls: Recognition of Sick Building Syndrome Has Laid the Foundation to Raise Toxic Tort Litigation to New Heights, 26 Texas Tech. L. Rev. 1041 (1995).

Thomas Icard Jr. and W.C. Wright, Sick Building Syndrome and Building-related Illness Claims: Defining the Practical and Legal Issues, 14 ABA Construction Lawyer 1 (Oct. 1994).

Mary Rose Kornreich, Minimizing Liability for Indoor Air Pollution, 4 Tulane Envtl. L. J. 61 (1990).

Arthur Larson, Workers Compensation Ch. 13 (1994 Supp.).

Michael S. Lieberman, B.J. DiMuro, J.B. Boyd, Multiple Chemical Sensitivity: An Emerging Area of Law, Trial 22 (July 1995).

Stephen A. Loewy, George W. Kelly, and Martha D. Nathanson, Indoor Pollution in Commercial Buildings: Legal Requirement and Emerging Trends, 3 U. Balt. J. Envtl. L. 29 (1993).

Norma L. Miller, *Healthy School Handbook* (National Education Association, 1995).

Wolfgang Preiser et al., *Post-Occupancy Evaluation* (Van Nostrand, 1988).

Michael Pyle, Environmental Law in an Office Building: The Sick Building Syndrome, 9 J. Envtl. L. & Litig. 173 (1994).

David Reisman, Strict Liability and Sick Building Syndrome: Defining a Building as a Product Under Restatement (Second) of Torts, Section 402A, 10 J. Nat. Resources & Envtl. L. 35 (1995).

Susan G. Rosmarin, An Indoor Air Litigation Primer, Indoor Pollution Law Report, 10 No. 12 Envtl. Compliance & Litigation Strategy 7 (May, 1995).

Kathleen Sablone, A Spark in the Battle Between Smokers and Nonsmokers: Johannesen v New York City Dept of Housing Preservation & Development, 36 Boston College L. Rev. 1089 (1996).

Howard Sandler, Multiple Chemical Sensitivity, Occupational Hazards 53 (April 1993).

Bradley Soos, Note, Adding Smoke to the Cloud of Tobacco Litigation—A New Plaintiff: The Involuntary Smoker, 23 Valparaiso U. L. Rev. 111 (1988).

Nina G. Stillman and John R. Wheeler, The Expansion of Occupational Safety & Health Law, 62 Notre Dame L. Rev. 969 (1987).

Abba Terr, Emil Bardana, and Leonard Altman, Idiopathic Environmental Intolerances (IEI), American Academy of Allergy, Asthma and Immunology Academy News 12 (June/July 1997).

Melissa A. Vallone, Employer Liability for Workplace Environmental Tobacco Smoke: Get Out of the Fog, 30 Valparaiso U. L. Rev. 811 (1996).

Federal Agencies with Roles in Addressing IEQ

Bonneville Power Administration
P.O. Box 3621-RMRD
Portland, OR 97208
503-230-5475
Electric utility supplier with extensive research on energy-conservation and residential-heating matters.

Consumer Product Safety Commission
5330 East-West Highway
Bethesda, MD 20207
800-638-CPSC
Reviews complaints regarding the safety of consumer products and takes action to ensure product safety.

General Services Administration
18th and F Streets, NW
Washington, DC 20405
202-501-1464
Writes indoor air quality policy for federal building assessments. Assesses complaints and provides remedial action.

U.S. Department of Energy Office of Conservation and Renewable Energy
1000 Independence Avenue, SW, CE-43
Washington, DC 20585
202-586-9455
Quantifies the relationship between energy conservation, adequate ventilation, and acceptable indoor air quality.

U.S. Department of Health and Human Services Office on Smoking and Health
National Center for Chronic Disease Prevention and Health Promotion, Centers for Disease Control
1600 Clifton Road, NE
Mail Stop K50
Atlanta, GA 30333
404-488-5705
Distributes information about the health effects of passive smoking and strategies for eliminating exposure to ETS.

Tennessee Valley Authority
Occupational Hygiene Department
328 Multipurpose Building
Muscle Shoals, AL 35660
205-386-2314
Provides building surveys and assessments associated with employee indoor air quality complaints.

Public Information Center
EPA IAQ Information Clearing
House
401 M Street, SW
Washington, DC 20460
202-260-2080
IAQ INFO: 800-438-4318 or 202-484-
1307
Distributes IAQ publications and pro-
vides information to the public on an
IAQ hotline.
(PM-211B)

National Pesticides Telecommunica-
tions
800-858-PEST (in Texas, 806-743-
3091)
Provides information on pesticides.

TSCA Hotline Service
202-554-1404
Provides information on asbestos and
other toxic substances.

Appendix C

EPA and OSHA Contact Information

EPA Regional Offices
Address inquiries to the contacts in the EPA Regional Offices at the following addresses (effective 1997).

EPA Region 1 (CT, ME, MA, NH, RI, VT)
John F. Kennedy Federal Building
Boston, MA 02203
617-565-3232 (indoor air)
617-565-4502 (radon)
617-565-3744 (asbestos)
617-565-3265 (air pollution)

EPA Region 2 (NJ, NY, PR, VI)
26 Federal Plaza
New York, NY 10278
212-264-4410 (radon)
212-264-4410 (indoor air)
212-264-6671 (asbestos)
212-264-6770 (air pollution)

EPA Region 3 (DE, DC, MD, PA, VA, WV)
841 Chesnut Building
Philadelphia, PA 19107
215-597-8322 (indoor air)
215-597-4084 (radon)
215-597-3160 (asbestos)
215-597-1970 (air pollution)

EPA Region 4 (AL, FL, GA, KY, MS, NC, SC, TN)
345 Courtland Street, NE
Atlanta, GA 30365
404-347-2864 (indoor air)
404-347-3907 (radon)
404-347-5014 (asbestos)
404-347-5014 (air pollution)

EPA Region 5 (IL, IN, MI, MN, OH, WI)
230 South Dearborn Street
Chicago, IL 60604
Region 5 Environmental Hotline:
1-800-572-2515 (IL)
1-800-621-8431 (IN, MI, MN, OH, WI)
312-886-7930 (calls from outside Region 5)

EPA Region 6 (AR, LA, NM, OK, TX)
1445 Ross Avenue
Dallas, TX 75202-2733
214-655-7223

EPA Region 7 (IA, KS, MO, NE)
726 Minnesota Avenue
Kansas City, KS 66101
913-551-7020

EPA Region 8 (CO, MT, ND, SD, UT, WY)
999 18th Street Suite 500
Denver, CO 80202-2405
303-293-1440 (indoor air)
303-293-0988 (radon)
303-293-1442 (asbestos)
303-294-7611 (air pollution)

EPA Region 9 (AZ, CA, HI, NV, AS, GU)
75 Hawthorne Street, A-1-1
San Francisco, CA 94105
415-744-1133 (indoor air)
415-744-1045 (radon)
415-744-1136 (asbestos)
415-744-1135 (air pollution)

EPA Region 10 (AK, ID, OR, WA)
1200 Sixth Avenue
Seattle, WA 98101
206-553-2589 (indoor air)
206-553-7299 (radon)
206-553-4762 (asbestos)
206-553-1757 (air pollution)

OSHA Regional Offices

OSHA Region 1 (CT, ME, MA, NH, RI, VT)
133 Portland Street, 1st Floor
Boston, MA 02114
617-565-7164

OSHA Region 2 (NJ, NY, PR, VI)
210 Varick Street, Room 670
New York, NY 10014
212-337-2376

OSHA Region 3 (DE, DC, MD, PA, VA, WV)
Gateway Building, Suite 2100
3535 Market Street
Philadelphia, PA 19104
215-596-1201

OSHA Region 4 (AL, FL, GA, KY, MS, NC, SC, TN)
1375 Peachtree Street
NE, Suite 587
Atlanta, GA 30367
404-347-3573

OSHA Region 5 (IL, IN, MI, MN, OH, WI)
230 South Dearborn Street, Room 3244
Chicago, IL 60604
312-353-2220

OSHA Region 6 (AR, LA, NM, OK, TX)
525 Griffin Street, Room 602
Dallas, TX 75202
214-767-4731

OSHA Region 7 (IA, KS, MO, NE)
911 Walnut Street, Room 406
Kansas City, MO 64106
816-426-5861

OSHA Region 8 (CO, MT, ND, SD, UT, WY)
Federal Building, Room 1576
1961 Stout Street
Denver, CO 80294
303-844-3061

OSHA Region 9 (AZ, CA, HI, NV, AS, GU)
71 Stevenson Street, 4th Floor
San Francisco, CA 94105
415-744-6570

OSHA Region 10 (AK, ID, OR, WA)
1111 Third Avenue, Suite 715
Seattle, WA 98101-3212
206-442-5930

Standards and Reference Organizations

Air Pollution Control Association (APCA)
PO Box 2861
Pittsburgh, PA 15230
412-232-3444

American Cancer Society
261 Madison Avenue
New York, NY 10016
212-599-3600

American Chemical Society
1155 16th Street, NW
Washington, DC 20036
202-872-4600

American Industrial Hygiene Association
475 Wolf Ledges Parkway
Akron, OH 44311-1087
216-762-7294

American Lung Association
1740 Broadway
New York, NY 10019-4374
212-315-8700

American Medical Association (AMA)
535 North Dearborn
Chicago, IL 60610
312-645-5000

American Society for Testing and Materials (ASTM)
1916 Race Street
Philadelphia, PA 19103
215-299-5400

American Society for Heating, Refrigerating, and Air Conditioning Engineers (ASHRAE)
1791 Tullie Circle, NE
Atlanta, GA 30329
404-636-8400

Asbestos Abatement Council
25 K Street, NE Suite 300
Washington, DC 20002
202-783-AWCI

Asbestos Information Association/North America
1745 Jefferson Davis Highway
Arlington, VA 22202
703-979-1150

Building Owners and Managers Association (BOMA) International
1250 Eye Street, NW, Suite 200
Washington, DC 20005
202-289-7000

Centers for Disease Control (CDC)
1600 Clifton Road
Atlanta, GA 30333
404-639-3534

Chemical Manufacturers Association
2501 M Street, NW
Washington, DC 20037
202-887-1100

Consumers Federation of America
1424 16th Street, NW, Suite 604
Washington, DC 20036
202-387-6121

Formaldehyde Institute
1330 Connecticut Avenue, NW, Suite 300
Washington, DC 20036
202-659-0060

National Academy of Sciences
2101 Constitution Avenue, NW
Washington, DC 20418
202-334-2000

National Coalition Against Misuse of Pesticides (NCAMP)
530 7th Street, SE
Washington, DC 20003
202-543-5450

National Institute of Building Sciences (NIBS)
1015 15th Street, NW
Washington, DC 20005
202-347-5710

National Pest Control Association
8100 Oak Street
Dunn Loring, VA 22027
703-573-8330

National Research Council
2101 Constitution Avenue, NW
Washington, DC 20418
202-334-2000

Safe Building Alliance
655 15th Street, NW, Suite 1200
Washington, DC 20005

Smoking Policy Institute
914 East Jefferson
Seattle, WA 98122
206-324-4444

The Tobacco Institute
1875 Eye Street, NW, Suite 800
Washington, DC 20006
202-457-4800

United States Environmental Protection Agency (EPA)
4101 M Street, SW
Washington, DC 20460
202-382-3949

United States Consumer Product Safety Commission
5330 East-West Highway
Bethesda, MD 20207
301-492-6554

Appendix *E*

ASHRAE Guidelines

ASHRAE Guidelines 1–1989. Guideline for the Commissioning of HVAC Systems. 1989.

ASHRAE Standard 52–76. Method of Testing Air-Cleaning Devices Used in General Ventilation for Removing Particulate Matter. 1976.

ASHRAE Standard 55–1981. Thermal Environmental Conditions for Human Occupancy. 1981.

ASHRAE Standard 62–1989. Ventilation for Acceptable Indoor Air Quality. 1989.

1. Internet, Encyclopedia Brittanica: http://www.eb.com:195/cgi-bin/g?DocF=macro/5005/71/115
2. Williams, Phillip, and Robert Greene. 1996. Indoor air quality investigation protocols. *J. Environmental Health* 6–13.

Private-Sector Contacts

Information on indoor air quality topics is available from the following private-sector organizations, among others.

BUILDING-MANAGEMENT ASSOCIATIONS

Association of Physical Plant Administrators of Universities and Colleges
1446 Duke Street
Alexandria, VA 22314-3492
703-684-1446

Building Owners and Managers Association International
1201 New York Avenue, NW, Suite 300
Washington, DC 20005
202-408-2684

Institute of Real Estate Management
430 North Michigan Avenue
Chicago, IL 60611
312-661-1930

International Council of Shopping Centers
1199 North Fairfax Street, Suite 204
Alexandria, VA 22314
703-549-7404

International Facilities Management Association
Summit Tower, Suite 1710
11 Greenway Plaza
Houston, TX 77046
713-623-4362

National Apartment Association
1111 14th Street, NW, Suite 900
Washington, DC 20005
202-842-4050

National Association of Industrial and Office Parks
1215 Jefferson Davis Highway, Suite 100
Arlington, VA 22202
703-979-3400

PROFESSIONAL AND STANDARD-SETTING ORGANIZATIONS

Air and Waste Management Association
P.O. Box 2861
Pittsburgh, PA 15230
412-232-3444

Air-Conditioning and Refrigeration Institute
1501 Wilson Boulevard, Suite 600
Arlington, VA 22209
703-524-8800

American Conference of Governmental Industrial Hygienists
6500 Glenway Avenue, Building D-7
Cincinnati, OH 45211
513-661-7881

American Industrial Hygiene Association
PO Box 8390
345 White Pond Drive
Akron, OH 44320
216-873-2442

American Society for Testing and Materials
1916 Race Street
Philadelphia, PA 19103
215-299-5571

American Society of Heating, Refrigerating, and Air-Conditioning Engineers
1791 Tullie Circle, NE
Atlanta, GA 30329
404-636-8400

National Conference of States on Building Codes and Standards, Inc.
505 Huntmar Park Drive, Suite 210
Herndon, VA 22070
703-437-0100

PRODUCT MANUFACTURERS

Adhesive and Sealant Council
1627 K Street, NW, Suite 1000
Washington, DC 20006-1707
202-452-1500

Asbestos Information Association
1745 Jefferson Davis Highway,
Room 509
Arlington, VA 22202
703-979-1150

Business Council on Indoor Air Quality
1225 19th Street, Suite 300
Washington, DC 20036
202-775-5887

Carpet and Rug Institute
310 Holiday Avenue
Dalton, GA 30720
404-278-3176

Chemical Specialties Manufacturers Association
1913 I Street, NW
Washington, DC 20006
202-872-8110

Formaldehyde Institute, Inc.
1330 Connecticut Avenue, NW
Washington, DC 20036
202-822-6757

Foundations of Wall and Ceiling Industries
1600 Cameron Street
Alexandria, VA 22314-2705
703-548-0374

Gas Research Institute
8600 West Bryn Mawr Avenue
Chicago, IL 60631
312-399-8304

National Paint and Coatings Association
1500 Rhode Island Avenue, NW
Washington, DC 20005
202-462-6272

Thermal Insulation Manufacturers Association Technical Services
Air Handling Committee
1420 King Street
Alexandria, VA 22314
703-684-0474

BUILDING SERVICE ASSOCIATIONS

Air-Conditioning and Refrigeration Institute
1501 Wilson Boulevard, 6th floor
Arlington, VA 22209
703-524-8800

Air-Conditioning Contractors of America
1513 16th Street, NW
Washington, DC 20036
202-483-9370

American Consulting Engineers Council
1015 15th Street, NW, Suite 802
Washington, DC 20005
202-347-7474

Associated Air Balance Council
1518 K Street, NW
Washington, DC 20005
202-737-0202

Association of Energy Engineers
4025 Pleasantdale Road, Suite 420
Atlanta, GA 30340
404-447-5083

Association of Specialists in Cleaning and Restoration International
10830 Annapolis Junction Road, Suite 312
Annapolis Junction, MD 20701
301-604-4411

National Air Duct Cleaners Association
1518 K Street, NW, Suite 503
Washington, DC 20005
202-737-2926

National Association of Power Engineers
3436 Haines Way, Suite 101
Falls Church, VA 22041
703-845-7055

National Energy Management Institute
601 North Fairfax Street, Suite 160
Alexandria, VA 22314
703-739-7100

National Environmental Balancing Bureau
1385 Piccard Drive
Rockville, MD 20850
301-977-3698

National Pest Control Association
8100 Oak Street
Dunn Loring, VA 22021
703-573-8330

Sheet Metal and Air Conditioning Contractors National Association
4201 LaFayette Center Drive
Chantilly, VA 22021
703-803-2980

UNIONS

AFL-CIO
Department of Occupational Safety and Health
815 16th Street, NW
Washington, DC 20006
202-637-5000

American Federation of Government Employees
80 F Street, NW
Washington, DC 20001
202-737-8700

American Federation of State, County, and Municipal Employees
1625 L Street, NW
Washington, DC 20036
202-429-1215

American Federation of Teachers
555 New Jersey Avenue, NW
Washington, DC 20001
202-879-4400

Communication Workers of America
501 3rd Street, NW
Washington, DC 20001
202-434-1160

International Union of Operating Engineers
1125 17th Street, NW
Washington, DC 20036
202-429-9100

Service Employees International Union
1313 L Street, NW
Washington, DC 20005

ENVIRONMENT/HEALTH/CONSUMER ORGANIZATIONS

American Academy of Allergy and Immunology
611 East Wells Street
Milwaukee, WI 53202
414-272-6071

American Lung Association
1740 Broadway
New York, NY 10019
(or your local lung association)

Consumer Federation of America
1424 16th Street, NW, Suite 604
Washington, DC 20036

National Center for Environmental Health Strategies
1100 Rural Avenue
Voorhees, NJ 08043
609-429-5358

National Environmental Health Association
720 South Colorado Boulevard
South Tower, Suite 970
Denver, CO 80222
303-756-9090

Appendix G

Publications

Items marked * are available from EPA Public Information Center (PM-211B), 401 M Street, SW, Washington, DC 20460, 202-260-2080.

Items marked ** are available from TSCA Assistance Hotline (TS-799), 401 M Street, SW, Washington, DC 20460, 202-554-1404.

Items marked *** are available from NIOSH Publications Dissemination, 4676 Columbia Parkway, Cincinnati, OH 45202, 513-533-8287.

BIBLIOGRAPHY

Sheet Metal and Air Conditioning Contractors National Association (SMACNA). *Indoor Air Quality.* 1988. 8224 Old Courthouse Road, Vienna, VA 22180.

U.S. Environmental Protection Agency and the Public Health Foundation. *Directory of State Indoor Air Contacts.* Updated, 1991.

U.S. Environmental Protection Agency. 1988. *Project Summaries: Indoor Air Quality in Public Buildings.* 1988. Contains findings of research on IAQ in 10 new public and commercial buildings and on building-material emissions.

U.S. Environmental Protection Agency and the U.S. Consumer Product Safety Commission. *The Inside Story: A Guide to Indoor Air Quality.* 1988. Addresses residential indoor air quality primarily, but contains a section on offices.

U.S. Environmental Protection Agency. *Sick Building Syndrome.* Indoor Air Quality Fact Sheet #4. Revised, 1991.

U.S. Environmental Protection Agency. *Ventilation and Air Quality in Offices.* Indoor Air Quality Fact Sheet #3. Revised, 1990.

World Health Organization. *Air Quality Guidelines for Europe.* 1987. WHO Regional Publications, European Series No. 23. Available from WHO Publications Center USA, 49 Sheridan Avenue, Albany, NY 12210.

ASBESTOS

A Guide to Monitoring Airborne Asbestos in Buildings. 1989. Environmental Sciences, Inc., 105 E. Speedway Blvd., Tucson, Arizona 85705.

U.S. Environmental Protection Agency. *A Guide to Respiratory Protection for the Asbestos Abatement Industry.* 1986. EPA 560/OTS 86-001.**

U.S. Environmental Protection Agency. *Abatement of Asbestos-Containing Pipe Insulation.* 1986. Technical Bulletin No. 1986-2.**

U.S. Environmental Protection Agency. *Asbestos Abatement Projects: Worker Protection.* Final Rule 40 CFR. 763. February 1987.**

U.S. Environmental Protection Agency. *Asbestos in Buildings: Guidance for Service and Maintenance Personnel* (in English and Spanish). 1985. EPA 560/5-85-018.**

U.S. Environmental Protection Agency. *Asbestos in Buildings: Simplified Sampling Scheme for Surfacing Materials.* 1985. 560/5-85-030A.**

U.S. Environmental Protection Agency. *Guidance for Controlling Asbestos-Containing Materials in Buildings.* 1985. EPA 560/5-85-024.**

U.S. Environmental Protection Agency. *Managing Asbestos in Place: A Building Owner's Guide to Operations and Maintenance Programs for Asbestos-Containing Materials.* 1990.**

U.S. Environmental Protection Agency. *Measuring Airborne Asbestos Following an Abatement Action.* 1985. EPA 600/4-85-049.**

U.S. Environmental Protection Agency. *National Emissions Standards for Hazardous Air Pollutants.* 40 Code of Federal Regulation 61. April 1984.**

U.S. Department of Labor. OSHA Regulations. *General Industry Asbestos Standard.* 29 Code of Federal Regulation 1910.1001.

U.S. Department of Labor/OSHA. *Construction Industry Asbestos Standard.* June 1986; amended September 1988. 29 Code of Federal Regulations 1926.58.

U.S. Department of Labor. OSHA Regulations. *Respiratory Protection Standard.* 29 Code of Federal Regulation 1910.134. June 1974. DOL-OSHA Docket, 200 Constitution Avenue, NW, Room N 2625, Washington, DC 20210.

BIOLOGICALS

American Council of Governmental Industrial Hygienists. *Guidelines for the Assessment of Bioaerosols in the Indoor Environment.* 1989. 6500 Glenway Avenue, Building D-7, Cincinnati, OH 45211.

Morey, P., J. Feeley, and J. Otten. *Biological Contaminants in Indoor Environments.* 1990. American Society for Testing and Materials Publications, 1916 Race Street, Philadelphia, PA 19103.

BUILDING MANAGEMENT, INVESTIGATION, AND REMEDIATION

Bazerghi, Hani, and Catherine Arnoult. *Practical Manual for Good Indoor Air Quality.* 1989. Quebec Association for Energy Management, 1259 Berri Street, Suite 510, Montreal, Quebec, Canada, H21 4C7.

Hansen, Shirley J. *Managing Indoor Air Quality.* 1991. Fairmount Press, 700 Indian Trail, Lilburn, GA 30247.

U.S. Department of Health and Human Services. Public Health Service. Centers for Disease Control. National Institute for Occupational Safety and Health. *Indoor Air Quality: Selected References.* 1989.

U.S. Department of Health and Human Services. Public Health Service. Centers for Disease Control. National Institute for Occupational Safety and Health. *Guidance for Indoor Air Quality Investigations.* 1987.

ENVIRONMENTAL TOBACCO SMOKE

National Research Council. *Environmental Tobacco Smoke: Measuring Exposures and Assessing Health Effects.* 1986. National Academy Press, 2001 Wisconsin Avenue, NW, Washington, DC 20418.

U.S. Department of Health and Human Services. Public Health Service. Office on Smoking and Health. *The Health Consequences of Involuntary Smoking, a Report of the Surgeon General.* 1986. 1600 Clifton Road, NE (Mail Stop K50), Atlanta, GA 30333.

U.S. Department of Health and Human Services. Public Health Service. Centers for Disease Control. National Institute for Occupational Safety and Health. *Current Intelligence Bulletin 54: Environmental Tobacco Smoke in the Workplace—Lung Cancer and Other Health Effects.* DHHS (NIOSH) Publication No. 91–108. 1991.***

U.S. Department of Health and Human Services. National Cancer Institute. Office of Cancer Communications. *A series of one-page information sheets on all aspects of smoking in the workplace.* For copies, call 800-4-CANCER.

U.S. Environmental Protection Agency. *Environmental Tobacco Smoke.* Indoor Air Quality Fact Sheet #5. 1989.*

PCBS

U.S. Department of Health and Human Services. Public Health Service. Centers for Disease Control. National Institute for Occupational Safety and Health. *Current Intelligence Bulletin 45: Polychlorinated Biphenyls—Potential*

Health Hazards from Electrical Equipment Fires or Failures. DHHS (NIOSH) Publication No. 86–111. 1977. Available from the National Technical Information Service, 5285 Port Royal Road, Springfield, VA 22161.

U.S. Environmental Protection Agency. *Transformers and the Risk of Fire: A Guide for Building Owners.* 1986. OPA/86-001.

RADON

U.S. Environmental Protection Agency. *State Proficiency Report.* 1991. EPA 520/1-91-014. Available from state radon offices and will provide a list of laboratories that have demonstrated competence in radon measurement analysis.

STANDARDS AND GUIDELINES

American Conference of Government Industrial Hygienists. *Threshold Limit Values and Biological Exposure Indices.* 1995–1996. 6500 Glenway Avenue, Building D-7, Cincinnati, OH 45211.

U.S. Department of Health and Human Services. Public Health Service. Centers for Disease Control. National Institute for Occupational Safety and Health. *NIOSH Recommendations for Occupational Safety and Health. Compendium of Policy Documents and Statements.* DHHS (NIOSH) Publication No. 91-109. 1991.**

U.S. Department of Labor. OSHA Regulations. 29 CFR Part 1910.1000. *OSHA Standards for Air Contaminants.* Available from the U.S. Government Printing Office, Washington, DC 20402, 202-783-3238. Additional health standards for some specific air contaminants are also available in Subpart Z.

TRAINING

American Industrial Hygiene Association (AIHA). P.O. Box 8390, 345 White Pond Drive, Akron, OH 44320, 216-873-2442. Sponsors indoor air quality courses in conjunction with meetings for AIHA members only.

American Society of Heating, Refrigerating, and Air-Conditioning Engineers (ASHRAE). 1791 Tullie Circle NE, Atlanta, GA 30329, 404-636-8400. Sponsors professional development seminars on indoor air quality.

NIOSH Division of Training and Manpower Development and NIOSH-funded Educational Resource Centers. 4676 Columbia Parkway, Cincinnati, OH 45226, 1-800-35-NIOSH. Provide training to occupational safety and health professionals and paraprofessionals.

OSHA Training Institute. 155 Times Drive, Des Plaines, IL 60018, 708-297-4913. Provides courses to assist health and safety professionals in evaluating indoor air quality.

Safe Drinking Water Act (SDWA) *Contact Information*

MAXIMUM CONTAMINANT LEVEL GOALS *AND MAXIMUM CONTAMINANT LEVELS*

EPA's Safe Drinking Water Hotline (SDWH) is available to help community, state, and local officials and the public understand the regulations and programs developed in response to the Safe Drinking Water Act amendments of 1986 and 1996. The SDWH provides information on EPA's drinking-water standards and monitoring requirements, contaminant-specific health advisories, and ground and source water protection efforts. Consumer information and referrals to local contacts and other sources of information are also available.

If you have additional questions, please contact EPA's Safe Drinking Water Hotline by telephone at 800-426-4791 (202-260-7908 outside the continental U.S.); by fax at 202-260-8072; or by e-mail at hotline-sdwa@epamail.epa.gov.

Additional information about EPA is available at URL=http://www.epa.gov. More information on EPA's Office of Water is available at URL=http://www .epa.gov/OWOW/.

Following is a list of drinking-water contaminants with maximum contaminant level goals (MCLGs) and maximum contaminant levels (MCLs).

Contaminant	MCLG[1]	MCL[1]
Acrylamide	zero	TT[2]
Alachlor (Lasso)	zero	0.002
Aldicarb[3]	0.001	0.003
Aldicarb sulfone[3]	0.001	0.002
Aldicarb sulfoxide[3]	0.001	0.004
Antimony	0.006	0.006
Arsenic	none[4]	0.05
Asbestos	7 MFL[5]	7 MFL[5]
Atrazine (Atranex, Crisazina)	0.003	0.003
Barium	2	2
Benzene	zero	0.005
Benzo[a]Pyrene (PAH)	zero	0.002
Beryllium	0.004	0.004
Beta particle and photon emitters	zero	4 mrem/yr
Cadmium	0.005	0.005
Carbofuran (Furadan 4F)	0.04	0.04
Carbon tetrachloride	zero	0.005
Chlordane	zero	0.002
Chromium	0.1	0.1
Total coliforms (Total)	zero	none
Copper	1.3	TT[6]
Cyanide	0.2	0.2
2,4-D (Formula 40, Weeder 64)	0.07	0.07
Dalapon	0.2	0.2
Di(2-ethylhexyl)adipate	0.4	0.4
Di(2-ethylhexyl)phthalate	zero	0.006
Dibromochloropropane (DBCP)	zero	0.0002
o-Dichlorobenzene	0.6	0.6
p-Dichlorobenzene	0.075	0.075
1,2-Dichloroethane	zero	0.005
1,1-Dichloroethylene	0.007	0.007
cis-1,2-Dichloroethylene	0.07	0.07
trans-1,2-Dichloroethylene	0.1	0.1

Dichloromethane	zero	0.005
1,2-Dichloropropane	zero	0.005
Dinoseb	0.007	0.007
Diquat	0.02	0.02
Endothall	0.1	0.1
Endrin	0.002	0.002
Epichlorohydrin	zero	TT[2]
Ethylbenzene	0.7	0.7
Ethylene dibromide	zero	0.00005
Fluoride	4.0	4.0
Giardia lamblia	zero	TT[7]
Glyphosate	0.7	0.7
Gross alpha particle activity	zero	15 pCi/l[8]
Heptachlor (H-34, Heptox)	zero	0.0004
Heptachlor expoxide	zero	0.0002
Heterotrophic plate count	zero	TT[7]
Hexachlorobenzene	zero	0.001
Hexachlorocyclopentadiene	0.05	0.05
Lead	zero	TT[6]
Legionella	zero	TT[7]
Lindane	0.0002	0.0002
Mercury	0.002	0.002
Methoxychlor (DMDT, Marlate)	0.04	0.04
Monochlorobenzene	0.1	0.1
Nickel	0.1[9]	0.1[9]
Nitrate (as nitrogen)	10	10
Nitrite (as nitrogen)	1	1
Nitrate/nitrite (Total)	10	10
Oxamyl (Vydate)	0.2	0.2
Pentachlorophenol	zero	0.0005
Picloram	0.5	0.5
Polychlorinated bipenyls (PCBs)	zero	0.001
Radium 226 & Radium 228 comb.	zero	5 pCi/l[8]
Selenium	0.05	0.05

Simazine	0.004	0.004
Styrene	0.1	0.1
Sulfate	deferred[10]	deferred[10]
2,3,7,8-TCDD (Dioxin)	zero	0.00000003
Tetrachloroethylene	zero	0.005
Thallium	0.0005	0.002
Toluene	1.0	1.0
Total trihalomethanes	—	0.10
Toxaphene	zero	0.003
2,4,5-TP (Silvex)	0.05	0.05
1,2,4-Trichlorobenzene	0.07	0.07
1,1,1-Trichloroethane	0.2	0.2
1,1,2-Trichloroethane	0.003	0.005
Trichloroethylene	zero	0.005
Turbidity	zero	TT[7]
Vinyl chloride	zero	0.002
Viruses	zero	TT[7]
Xylenes	10.0	10.0

[1] Unless otherwise specified, the unit of measurement is milligrams per liter (mg/l).

[2] Each public water system must certify annually in writing to the state (using third-party or manufacturer's certification) that when acrylamide and epichlorohydrin are used in drinking water systems, the combination (or product) of dose and monomer level does not exceed the levels specified as follows: (1) Acrylamide = 0.05% dosed at 1 ppm (or equivalent) and (2) epichlorohydrin = 0.01% dosed at 20 ppm (or equivalent).

[3] These levels are not in effect. EPA plans to repropose levels in the future.

[4] The MCLG for this contaminant was withdrawn and is currently under review.

[5] MFL is million fibers per liter.

[6] The lead and copper rule is a treatement technique that requires water systems to take tap water samples from homes with lead pipes or copper pipes with lead solder and/or with lead service lines. If more than 10 percent of these samples exceed an action level of 1.3 mg/l for copper or 0.015 mg/l for lead, the system is triggered into additional treatment.

[7] These contaminants are regulated under the Surface Water Treatement Rule (40 CFR 141.70–141.75).

[8] pCi/l is picocuries per liter.

[9] The MCLG and MCL for nickel have been withdrawn.

[10] The proposed MCLG and MCL for sulfate is 500 mg/l.

Considerations When Responding to Health Complaints

1. Determine whether the complaint is sudden and represents a potential true emergency.
2. If complaints are chronic, be certain to hire consultants who are knowledgeable about health issues and credible to employees.
3. If levels of distress are high, indicating an emotionally charged issue, be sure that those dealing with the workers are prepared to manage the psychology of the situation.
4. Health surveys or questionnaires can be disastrous if not properly developed and carefully used. In these investigations it is far preferable to conduct individual interviews.
5. Never implement expensive solutions without proof that the identified problem is the actual cause of health complaints or alleged risks.
6. Be certain that your consultants are capable of communicating their findings and addressing worker concerns. Concerns, distress, and perceptions are as important as the realities of potential contamination.

RULES OF THUMB REGARDING HEALTH EFFECTS
AND THE OFFICE ENVIRONMENT

1. The wider the variety of symptoms, the less likely that there is a common or single environmental cause.
2. Most symptoms that people associate with the workplace are not caused by bad air.

3. Discomfort and ill health are not necessarily the same thing; workers may be uncomfortable but not sick.
4. When physical findings (e.g., rash, eye irritation) accompany symptoms, a diagnosis and causal attribution are easier to make.
5. An environmental finding, such as an unbalanced HVAC system, dirty filters, or mold, does not necessarily explain reported symptoms.

Appendix J

Research Report to Building Owners and Managers Association International, Institute of Real Estate Management, International Council of Shopping Centers, National Apartment Association, National Association of Realtors®, and National Multihousing Council (Results of a National Survey of Workplace Professionals, August 24, 1995)

BACKGROUND

Several organizations in the building ownership and management community have come together to underwrite this study about the perceived quality of air in the American professional workplace. The study was specifically designed to emulate the distribution of buildings documented in the Energy Information Administration's *Commercial Buildings Characteristics 1992* study. In the face of possible stepped-up federal intervention in building operation and maintenance matters, this study was designed to provide an up-to-date and accurate reflection of the perceived quality of building air quality among those who know it best—namely, the persons who work in these buildings.

RESEARCH OBJECTIVES

The objectives of this study included the following:

- Measure the current level of satisfaction (or dissatisfaction) with building air quality among workplace professionals.
- Identify and measure the elements of dissatisfaction among those who believe the quality of air in their workplace is a problem.
- Assess the extent to which perceived workplace air quality problems contribute to upper respiratory ailments and severe headaches.

METHODOLOGY

Interviewing for this survey took place between June 27 and August 1, 1995. All interviewing was done by the trained professional interviewers of Southeastern Institute of Research (SIR) at SIR's central telephone interviewing facility in Richmond, Virginia, where all work was carefully supervised and systematically reviewed for consistency, accuracy, and quality of logic.

A random sample of workplace professionals was used for this survey. The survey sample was designed to replicate the geographic distribution of the U.S. Department of Energy, Energy Information Administration's *Commercial Buildings Characteristics 1992* survey. Specifically, the sampling geography of this survey used the same nine-region geographic distribution as was used in the 1992 DOE survey. Respondents were interviewed at home.

A standardized survey questionnaire, developed based on input from the real estate organizations involved and SIR, was used. A copy of the final questionnaire is included as an appendix to this document.

We are not aware of any unusual circumstances in the marketplace at the time of this survey which might have been expected to unduly influence the outcome of the survey.

SUMMARY OF THE FINDINGS

This survey of eight hundred and fifty-eight persons who work in commercial buildings throughout the United States reveals interesting insights into the extent to which building air quality is perceived to be a problem.

First and foremost, this survey finds that building air quality is *not* perceived to be a problem for most workplace professionals. Asked to assess the quality of air in their buildings, the vast majority of respondents gave their buildings positive ratings. Specifically, eight-in-ten (80%) survey respondents voiced the opinion that the quality of the air where they work is either "O.K." or better; eleven percent believe the quality of air where they work is fair; and nine percent believe the quality of the air where they work is poor.

Asked more specifically whether building air quality is a problem, over eight-in-ten (85%) respondents confirmed their earlier response by answering "never" or "once in a while."

The percentage of respondents who report that building air quality is a problem "often" or more frequently represents a small minority—only fifteen percent—of respondents overall. The percentage who describe building air quality problems as ongoing account for an even smaller 6% of the entire sample of respondents.

We find it interesting that among the small number of respondents who report any concerns about their building air quality, the majority of these concerns relate more to *comfort* factors—i.e. too hot/too cold—rather than issues involving air pollution. We further believe it is worthy of mention that the second largest class of perceived building air quality problems, such as food and photocopier odors, are occupant related.

This survey does not present evidence of a causal relationship between workplace air quality and factors such as job satisfaction and exposure to stress.

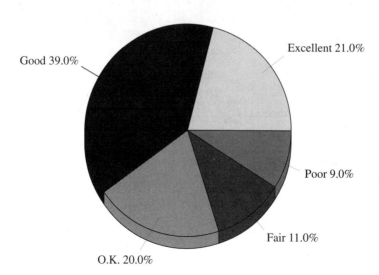

Figure J.1.   Rating of workplace air quality.

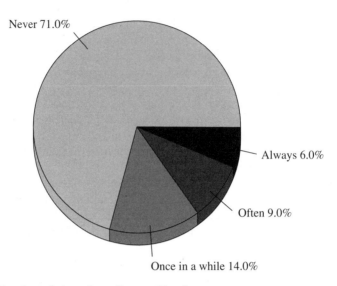

Figure J.2. Is workplace air quality a problem?

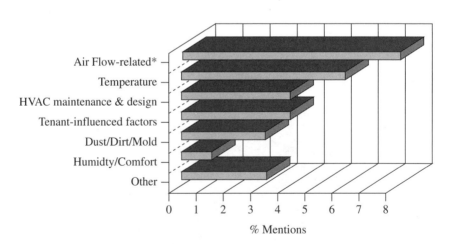

"Air flow-related" includes complaints pertaining to uneven air distribution, too much air flow, as well as too little air flow. It is important to note the "air flow-related" does not necessarily mean introduction of insufficient levels of outdoor air brought indoors.

Figure J.3. Perceived causes of workplace air problems.

However, to the extent that there are at least correlations between these factors, the survey shows that:

- Persons who are satisfied with their jobs are less likely to mention problems with the quality of the air in their workplaces.
- Conversely, many people who are dissatisfied with their jobs are more likely to be dissatisfied with the quality of the air in their workplaces.
- Persons who profess to experience job stress all or most of the time are more likely to express dissatisfaction with the quality of the air in their workplaces.
- Persons who complain about workplace air quality are also more likely to say that their workplace lighting is either too bright or too dim compared to those who do not believe their workplaces have air quality problems.

In finding that building air quality problems are a matter of importance to a small segment of respondents overall, the survey further documents that problems perceived to be related to building air quality are problematic to an even smaller segment of workplace professionals. Specifically, fewer than one-in-ten (8%) of respondents say they have *ever* been made ill by what they perceived to be workplace air quality problems. Fewer still—only 3%—say they have lost time from work during the past year due to what they perceive to be building air quality problems.

Among those very few who have been ill due to what they believe are workplace air quality problems, the most frequently mentioned of a wide array of maladies are conditions that can as easily be attributed to home and environmental factors as to the quality of workplace air. Indeed, many of the symptoms mentioned are in fact associated with general conditions such as common colds, allergy-related conditions, chronic fatigue syndrome, and other at-large conditions.

Other patterns that appear in the data include:

- The taller one's building, the more likely one is to be unhappy with workplace air quality. The median number of floors in buildings of persons who believe the air quality in their workplaces is a problem is, at 3.2 floors, nearly double that of persons who do not believe the quality of air in their workplaces is satisfactory.
- There is a slightly greater likelihood of mentioning workplace air quality problems among those working in older buildings.
- The likelihood of complaining about workplace air quality is also greater among those who work in modular and open work spaces than among those who work in work spaces that have floor-to-ceiling walls.
- There appears to be only a slightly greater mention of workplace air quality concerns about those in buildings where windows are sealed compared to those in buildings where windows can be opened.

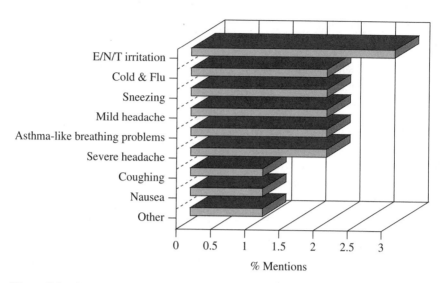

Figure J.4. Symptoms of illness from perceived poor air quality (among all survey respondents).

Persons in clerical positions and those in government occupations were more likely to mention building air quality problems in this survey. Conversely, professional/managerial workers and persons working for non-government employers were more likely to perceive there to be no problem with workplace air quality.

On a geographic basis, this survey finds that perceived workplace air quality problems are more widely reported in the Eastern region of the country. Survey respondents in the Middle Atlantic and South Atlantic, particularly those in Pennsylvania, Virginia, and the District of Columbia are noticeably over-represented among those complaining about building air quality. Conversely, respondents in the East and West South Central and Pacific regions are under-represented among those who complained about building air quality.

A surprisingly large percentage (37%) of those who identified air-quality problems in their buildings have chosen not to report their concerns. Only about two-thirds of the persons who believe the air in their buildings is a problem say they have ever complained to a person of authority. Persons who mentioned *comfort* issues—particularly temperature problems—appear to be the most likely to complain.

Complaints about building air quality are concentrated most among persons between the ages of 25 and 44, with noticeably lower levels of mentions among persons in other age groups. Women are more likely to be concerned about workplace air quality than men.

Among those who have complained about the quality of workplace air, this survey makes it clear that the majority of complaints were addressed promptly and professionally. What's more, persons who report having been made ill by what they perceive to be building air quality problems are *twice as likely* to say their complaints were handled promptly and professionally.

In summary, we believe this survey reliably documents that genuine building air quality problems are a much smaller problem than previously supposed. Furthermore, the survey documents that complaints about building air quality, when they are made, are generally handled promptly and professionally. The survey does not pretend to represent that there are no building air quality problems. But it does put these problems in the proper perspective.

Checklist for IEQ Evaluation of a Building

INDOOR AIR QUALITY FORMS

This section of the document is a collection of the forms that appear or are mentioned in the text. Consider making copies of the forms, blocking out the page information at the top of the copies, and then reproducing these copies for use in your building. Some or all of them may require adaptation to meet your specific needs. Blank formatted sheets are included for preparing your own HVAC Checklist and Pollutant and Source Inventory.

The forms appear in the following sequence:

IAQ Management Checklist (4 pages): for keeping track of the elements of the IAQ profile and IAQ management plan

Pollutant Pathway Record for IAQ Profiles: for identifying areas in which negative or positive pressures should be maintained.

Zone/Room Record: for recording information on a room-by-room basis on the topics of room use, ventilation, and occupant population

Ventilation Worksheet: to be used in conjunction with the Zone/Room Record when calculating quantities of outdoor air that are being supplied to individual zones or rooms

Source: Building Air Quality: A Guide for Building Owners and Facility Managers, U.S. Environmental Protection Agency and U.S. National Institute for Occupational Safety & Health, EPA/400/1-91/033, DHHS (NIOSH) 91-114, December 1991. (Federal publication, not copyrighted.)

IAQ Complaint Form: to be filled out by the complainant or by a staff person who receives information from the complainant

Incident Log: for keeping tract of each IAQ complaint or problem and how it is handled

Occupant Interview (2 pages): for recording the observations of building occupants in relation to their symptoms and conditions in the building

Occupant Diary: for recording incidents of symptoms and associated observations as they occur

Log of Activities and System Operation: for recording activities and equipment operating schedules as they occur

HVAC Checklist—Short Form (4 pages): to be used as a short form for investigating an IAQ problem, or for periodic inspections of the HVAC system. Duplicate pages 2 through 4 for each large air handling unit.

HVAC Checklist—Long Form (14 pages, followed by one blank formatted sheet): to be used for detailed inspections of the HVAC system or as a long form for investigating an IAQ problem. Duplicate pages 1 through 11 for each large air handling unit.

Pollutant Pathway Form For Investigations: to be used in conjunction with a floor plan of the building

Pollutant and Source Inventory (6 pages, followed by one blank formatted sheet): to be used as a general checklist of potential indoor and outdoor pollutant sources

Chemical Inventory: for recording information about chemicals stored or used within the building

Hypothesis Form (2 pages): to be used for summarizing what has been learned during the building investigation; a tool to help the investigator collect his or her thoughts

IAQ Management Checklist

Building Name: _____ Date: _____

Address: _____

Completed by (name/title): _____

Use this checklist to make sure that you have included all necessary elements in your IAQ profile and
IAQ management plan. *Sections 4 and 5* discuss the development of the IAQ profile and IAQ management plan.

Item	Date begun or completed (as applicable)	Responsible person (name, telephone)	Location ("NA" if the item is not applicable to this building)
IAQ PROFILE			
Collect and Review Existing Records			
HVAC design data, operating instructions, and manuals			
HVAC maintenance and calibration records, testing and balancing reports			
Inventory of locations where occupancy, equipment, or building use has changed			
Inventory of complaint locations			
Conduct a Walkthrough Inspection of the Building			
List of responsible staff and/or contractors, evidence of training, and job descriptions			
Identification of areas where positive or negative pressure should be maintained			
Record of locations that need monitoring or correction			
Collect Detailed Information			
Inventory of HVAC system components needing repair, adjustment, or replacement			
Record of control settings and operating schedules			

IAQ Management Checklist

Item	Date begun or completed (as applicable)	Responsible person (name, telephone)	Location ("NA" if the item is not applicable to this building)
Plan showing airflow directions or pressure differentials in significant areas			
Inventory of significant pollutant sources and their locations			
MSDSs for supplies and hazardous substances that are stored or used in the building			
Zone/Room Record			
IAQ MANAGEMENT PLAN			
Select IAQ Manager			
Review IAQ Profile			
Assign Staff Responsibilities/ Train Staff			
Facilities Operation and Maintenance			
■ confirm that equipment operating schedules are appropriate			
■ confirm appropriate pressure relationships between building usage areas			
■ compare ventilation quantities to design, codes, and ASHRAE 62-1989			
■ schedule equipment inspections per preventive maintenance plan or recommended maintenance schedule			
■ modify and use HVAC Checklist(s); update as equipment is added, removed, or replaced			
■ schedule maintenance activities to avoid creating IAQ problems			

IAQ Management Checklist

Item	Date begun or completed (as applicable)	Responsible person (name, telephone)	Location ("NA" if the item is not applicable to this building)
■ review MSDSs for supplies; request additional information as needed			
■ consider using alarms or other devices to signal need for HVAC maintenance (e.g., clogged filters)			
Housekeeping			
■ evaluate cleaning schedules and procedures; modify if necessary			
■ review MSDSs for products in use; buy different products if necessary			
■ confirm proper use and storage of materials			
■ review trash disposal procedures; modify if necessary			
Shipping and Receiving			
■ review loading dock procedures (*Note:* If air intake is located nearby, take precautions to prevent intake of exhaust fumes.)			
■ check pressure relationships around loading dock			
Pest Control			
■ consider adopting IPM methods			
■ obtain and review MSDSs; review handling and storage			
■ review pest control schedules and procedures			
■ review ventilation used during pesticide application			

IAQ Management Checklist

Item	Date begun or completed (as applicable)	Responsible person (name, telephone)	Location ("NA" if the item is not applicable to this building)
Occupant Relations			
▪ establish health and safety committee or joint tenant/ management IAQ task force			
▪ review procedures for responding to complaints; modify if necessary			
▪ review lease provisions; modify if necessary			
Renovation, Redecorating, Remodeling			
▪ discuss IAQ concerns with architects, engineers, contractors, and other professionals			
▪ obtain MSDSs; use materials and procedures that minimize IAQ problems			
▪ schedule work to minimize IAQ problems			
▪ arrange ventilation to isolate work areas			
▪ use installation procedures that minimize emissions from new furnishings			
Smoking			
▪ eliminate smoking in the building			
▪ if smoking areas are designated, provide adequate ventilation and maintain under negative pressure			
▪ work with occupants to develop appropriate non-smoking policies, including implementation of smoking cessation programs			

Pollutant Pathway Record For IAQ Profiles

This form should be used in combination with a floor plan such as a fire evacuation plan.

Building Name: _____ File Number: _____

Address: _____

Completed by: _____ Title: _____ Date: _____

Sections 2, 4 and 6 discuss pollutant pathways and driving forces.

Building areas that contain contaminant sources (e.g., bathrooms, food preparation areas, smoking lounges, print rooms, and art rooms) should be maintained under negative pressure relative to surrounding areas. Building areas that need to be protected from the infiltration of contaminants (e.g., hallways in multi-family dwellings, computer rooms, and lobbies) should be maintained under positive pressure relative to the outdoors and relative to surrounding areas.

List the building areas in which pressure relationships should be controlled. As you inspect the building, put a Y or N in the "Needs Attention" column to show whether the desired air pressure relationship is present. Mark the floor plan with arrows, plus signs (+) and minus signs (-) to show the airflow patterns you observe using chemical smoke or a micromanometer.

Building areas that appear isolated from each other may be connected by airflow passages such as air distribution zones, utility tunnels or chases, party walls, spaces above suspended ceilings (whether or not those spaces are serving as air plenums), elevator shafts, and crawlspaces. If you are aware of pathways connecting the room to identified pollutant sources (e.g., items of equipment, chemical storage areas, bathrooms), it may be helpful to record them in the "Comments" column, on the floor plan, or both.

Building Area (zone, room)	Use	Intended Pressure		Needs Attention? (Y/N)	Comments
		Positive (+)	Negative (-)		

Zone/Room Record

Building Name: _____ File Number: _____ Date: _____

Address: _____ Completed by: _____ Title: _____

This form is to be used differently depending on whether the goal is to *prevent* or to *diagnose* IAQ problems. During the development of a profile, this form should be used to record more general information about the entire building; during an investigation, the form should be used to record more detailed information about the complaint area and areas surrounding the complaint area or connected to it by pathways.

Use the last three columns when underventilation is suspected. Use the **Ventilation Worksheet** and *Appendix A* to estimate outdoor air quantities. Compare results to the design specifications, applicable building codes, or ventilation guidelines such as ASHRAE 62-1989. (See *Appendix A* for some outdoor air quantities required by ASHRAE 62-1989.) *Note:* For VAV systems, minimum outdoor air under reduced flow conditions must be considered.

Building Area (Zone/Room)	Use**	Source of Outdoor Air*	Mechanical Exhaust? (Write "No" or estimate cfm airflow)	Comments	Peak Number of Occupants or Sq. Ft. Floor Area**	Total Air Supplied (in cfm)***	Outdoor Air Supplied per Person or per 150 Sq. Ft. Area (in cfm)****

PROFILE AND DIAGNOSIS INFORMATION — DIAGNOSIS INFORMATION ONLY

* Sources might include air handling unit (e.g., AHU-4), operable windows, transfer from corridors
** Underline the information in this column if current use or number of occupants is different from design specifications
*** Mark the information with a **P** if it comes from the mechanical plans or an **M** if it comes from the actual measurements, such as recent test and balance reports.
**** ASHRAE 62-1989 gives ventilation guidance per 150 sq. ft.

Ventilation Worksheet

Building Name: _____ File Number: _____

Address: _____

Completed by (name): _____ Date: _____

This worksheet is designed for use with the **Zone/Room Record.** *Appendix A* provides guidance on methods of estimating the amount of ventilation (outdoor) air being introduced by a particular air handling unit. *Appendix B* discusses the ventilation recommendations of ASHRAE Standard 62-1989, which was developed for the purpose of preventing indoor air quality problems. Formulas are given below for calculating outdoor air quantities using thermal or CO_2 information.

The equation for calculating outdoor air quantities **using thermal measurements** is:

$$\text{Outdoor air (in percent)} = \frac{T_{return\ air} - T_{mixed\ air}}{T_{return\ air} - T_{outdoor\ air}} \times 100$$

Where: T = temperature in degrees Fahrenheit

The equation for calculating outdoor quantities **using carbon dioxide measurements** is:

$$\text{Outdoor air (in percent)} = \frac{C_s - C_r}{C_0 - C_r} \times 100$$

Where: C_s = ppm of carbon dioxide in the supply air (if measured in a room), or
C_s = ppm of carbon dioxide in the mixed air (if measured at an air handler)
C_r = ppm of carbon dioxide in the return air
C_o = ppm of carbon dioxide in the outdoor air

Use the table below to estimate the ventilation rate in any room or zone. *Note:* ASHRAE 62-1989 generally states ventilation (outdoor air) requirements on an occupancy basis; for a few types of spaces, however, requirements are given on a floor area basis. Therefore, this table provides a process of calculating ventilation (outdoor air) on either an occupancy or floor area basis.

Zone/Room	Percent of Outdoor Air	Total Air Supplied to Zone/Room (cfm)	Peak Occupancy (number of people) or Floor Area (square feet)	$D = \frac{B}{C}$ Total Air Supplied Per Person (or per square foot area)	$E = (A \times 100) \times D$ Outdoor air Supplied Per Person (or per square foot area)
	A	**B**	**C**	**D**	**E**

Indoor Air Quality Complaint Form

This form can be filled out by the building occupant or by a member of the building staff.

Occupant Name: _____ Date: _____

Department/Location in Building: _____ Phone: _____

Completed by: _____ Title: _____ Phone: _____

This form should be used if your complaint may be related to indoor air quality. Indoor air quality problems include concerns with temperature control, ventilation, and air pollutants. Your observations can help to resolve the problem as quickly as possible. Please use the space below to describe the nature of the complaint and any potential causes.

We may need to contact you to discuss your complaint. What is the best time to reach you? _____

So that we can respond promptly, please return this form to: _____

IAQ Manager or Contact Person

_____

Room, Building, Mail Code

OFFICE USE ONLY

File Number: _____ Received By: _____ Date Received: _____

Incident Log

Building Name: _____

Address: _____

Completed by (name): _____

Dates (from): _____ (to): _____

| File Number | Date | Problem Location | Investigation Record (check the forms that were used) | | | | | | | | | Outcome/Comments (use more than one line if needed) | Log Entry By (initials) |
			Complaint Form	Occupant Interview	Occupant Diary	Log of Activities	Zone/Room Record	HVAC Checklist	Pollutant Pathway	Source Inventory	Hypothesis Form		

Occupant Interview

Building Name: _____ File Number: _____

Address: _____

Occupant Name: _____ Work Location: _____

Completed by: _____ Title: _____ Date: _____

Section 4 discusses collecting and interpreting information from occupants.

SYMPTOM PATTERNS

What kind of symptoms or discomfort are you experiencing?

Are you aware of other people with similar symptoms or concerns? Yes _____ No _____

If so, what are their names and locations? _____

_____

Do you have any health conditions that may make you particularly susceptible to environmental problems?

❏ contact lenses ❏ chronic cardiovascular disease ❏ undergoing chemotherapy or radiation therapy

❏ allergies ❏ chronic respiratory disease ❏ immune system suppressed by disease or
other causes

❏ chronic neurological problems

TIMING PATTERNS

When did your symptoms start?

When are they generally worst?

Do they go away? If so, when?

Have you noticed any other events (such as weather events, temperature or humidity changes, or activities in the building) that tend to occur around the same time as your symptoms?

Occupant Interview

SPATIAL PATTERNS

Where are you when you experience symptoms or discomfort?

Where do you spend most of your time in the building?

ADDITIONAL INFORMATION

Do you have any observations about building conditions that might need attention or might help explain your symptoms (e.g., temperature, humidity, drafts, stagnant air, odors)?

Have you sought medical attention for your symptoms?

Do you have any other comments?

Occupant Diary

Occupant Name: _____ Title: _____ Phone: _____

Location: _____ File Number :_____

On the form below, please record each occasion when you experience a symptom of ill-health or discomfort that you think may be linked to an environmental condition in this building.

It is important that you record the time and date and your location within the building as accurately as possible, because that will help to identify conditions (e.g., equipment operation) that may be associated with your problem. Also, please try to describe the severity of your symptoms (e.g., mild, severe) and their duration (the length of time that they persist). Any other observations that you think may help in identifying the cause of the problem should be noted in the "Comments" column. Feel free to attach additional pages or use more than one line for each event if you need more room to record your observations.

Section 6 discusses collecting and interpreting occupant information.

Time/Date	Location	Symptom	Severity/Duration	Comments

Log of Activities and System Operation

Building Name: _____ Address: _____ File Number : _____

Completed by: _____ Title: _____ Phone: _____

On the form below, please record your observations of the HVAC system operation, maintenance activities, and any other information that you think might be helpful in identifying the cause of IAQ complaints in this building. Please report any other observations (e.g., weather, other associated events) that you think may be important as well.

Feel free to attach additional pages or use more than one line for each event.

Equipment and activities of particular interest:

Air Handler(s): _____

Exhaust Fan(s): _____

Other Equipment or Activities: _____

Date/Time	Day of Week	Equipment Item/Activity	Observations/Comments

HVAC Checklist - Short Form

Building Name: _____ Address: _____

Completed by: _____ Date: _____ File Number: _____

Sections 2, 4 and 6 and Appendix B discuss the relationships between the HVAC system and indoor air quality.

MECHANICAL ROOM

■ Clean and dry? _____ Stored refuse or chemicals? _____

■ Describe items in need of attention _____

MAJOR MECHANICAL EQUIPMENT

■ Preventive maintenance (PM) plan in use? _____

Control System

■ Type _____

■ System operation _____

■ Date of last calibration _____

Boiler

■ Rated Btu input _____ Condition _____

■ Combustion air: is there at least one square inch free area per 2,000 Btu input? _____

■ Fuel or combustion odors _____

Cooling Tower

■ Clean? no leaks or overflow? _____ Slime or algae growth? _____

■ Eliminator performance _____

■ Biocide treatment working? (list type of biocide) _____

■ Spill containment plan implemented? _____ Dirt separator working? _____

Chillers

■ Refrigerant leaks? _____

■ Evidence of condensation problems? _____

■ Waste oil and refrigerant properly stored and disposed of? _____

HVAC Checklist - Short Form   *Page 2 of 4*

Building Name: _____ Address: _____

Completed by: _____ Date: _____ File Number: _____

AIR HANDLING UNIT

■ Unit identification _____ Area served _____

Outdoor Air Intake, Mixing Plenum, and Dampers

■ Outdoor air intake location _____

■ Nearby contaminant sources? (describe)_____

■ Bird screen in place and unobstructed? _____

■ Design total cfm _____ outdoor air (O.A.) cfm _____ date last tested and balanced _____

■ Minimum % O.A. (damper setting) _____ Minimum cfm O.A. $\frac{\text{(total cfm x minimum \% O.A.)}}{100}$ = _____

■ Current O.A. damper setting (date, time, and HVAC operating mode) _____

■ Damper control sequence (describe) _____

■ Condition of dampers and controls (note date) _____

Fans

■ Control sequence _____

■ Condition (note date) _____

■ Indicated temperatures   supply air _____ mixed air _____ return air _____ outdoor air _____

■ Actual temperatures   supply air _____ mixed air _____ return air _____ outdoor air _____

Coils

■ Heating fluid discharge temperature _____ ΔT _____ cooling fluid discharge temperature _____ ΔT _____

■ Controls (describe) _____

■ Condition (note date) _____

Humidifier

■ Type _____ If biocide is used, note type _____

■ Condition (no overflow, drains trapped, all nozzles working?) _____

■ No slime, visible growth, or mineral deposits? _____

HVAC Checklist - Short Form

Building Name: _____ Address: _____

Completed by: _____ Date: _____ File Number: _____

DISTRIBUTION SYSTEM

Zone/ Room	System Type	Supply Air		Return Air		Power Exhaust		
		ducted/ unducted	cfm	ducted/ unducted	cfm	cfm	control	serves (e.g. toilet)

Condition of distribution system and terminal equipment (note locations of problems)

- Adequate access for maintenance? _____

- Ducts and coils clean and obstructed? _____

- Air paths unobstructed?   supply _____ return _____ transfer _____ exhaust _____ make-up _____

- Note locations of blocked air paths, diffusers, or grilles _____

- Any unintentional openings into plenums? _____

- Controls operating properly? _____

- Air volume correct? _____

- Drain pans clean? Any visible growth or odors? _____

Filters

Location	Type/Rating	Size	Date Last Changed	Condition (give date)

HVAC Checklist - Short Form

Building Name: _____ Address: _____

Completed by: _____ Date: _____ File Number: _____

OCCUPIED SPACE

Thermostat types ——————————————————————————————————————

Zone/ Room	Thermostat Location	What Does Thermostat Control? (e.g., radiator, AHU-3)	Setpoints		Measured Temperature	Day/ Time
			Summer	Winter		

Humidistat/Dehumidistat types ——————————————————————————————

Zone/ Room	Humidistat/ Dehumidistat Location	What Does It Control?	Setpoints (%RH)	Measured Temperature	Day/ Time

■ Potential problems (note location) _____

■ Thermal comfort or air circulation problems (drafts, obstructed airflow, stagnant air, overcrowding, poor thermostat location)

_____

_____

■ Malfunctioning equipment ——————————————————————————————

■ Major sources of odors or contaminants (e.g., poor sanitation, incompatible uses of space)

_____

HVAC Checklist - Long Form

Building: _____ File Number: _____

Completed by: _____ Title: _____ Date Checked: _____

Appendix B discusses HVAC system components in relation to indoor air quality.

Component	OK	Needs Attention	Not Applicable	Comments
Outside Air Intake				
Location _____ _____				
Open during occupied hours?				
Unobstructed?				
Standing water, bird droppings in vicinity?				
Odors from outdoors? (describe) _____ _____				
Carryover of exhaust heat?				
Cooling tower within 25 feet?				
Exhaust outlet within 25 feet?				
Trash compactor within 25 feet?				
Near parking facility, busy road, loading dock?				
Bird Screen				
Unobstructed?				
General condition?				
Size of mesh? ($1/2$" minimum)				
Outside Air Dampers				
Operation acceptable?				
Seal when closed?				

HVAC Checklist - Long Form

Building: _____ File Number: _____

Completed by: _____ Title: _____ Date Checked: _____

Component	OK	Needs Attention	Not Applicable	Comments
Actuators operational?				
Outdoor Air (O.A.) Quantity *(Check against applicable codes and ASHRAE 62-1989.)*				
Minimum % O.A. ————				
Measured % O.A. ———— *Note day, time, HVAC operating mode under "Comments"*				
Maximum % O.A. ————				
Is minimum O.A. a separate damper?				
For VAV systems: is O.A. increased as total system air-flow is reduced?				
Mixing Plenum				
Clean?				
Floor drain trapped?				
Airtightness				
■ of outside air dampers				
■ of return air dampers				
■ of exhaust air dampers				
All damper motors connected?				
All damper motors operational?				
Air mixers or opposed blades?				

HVAC Checklist - Long Form

Building: _____ File Number: _____

Completed by: _____ Title: _____ Date Checked: _____

Component	OK	Needs Attention	Not Applicable	Comments
Mixed air temperature control setting _____ °F				
Freeze stat setting _____ °F				
Is mixing plenum under negative pressure? *Note: If it is under positive pressure, outdoor air may not be entering.*				
Filters				
Type _____				
Complete coverage? (i.e., no bypassing)				
Correct pressure drop? *(Compare to manufacturer's recommendations.)*				
Contaminants visible?				
Odor noticeable?				
Spray Humidifiers or Air Washers				
Humidifier type				
All nozzles working?				
Complete coil coverage?				
Pans clean, no overflow?				
Drains trapped?				
Biocide treatment working? *Note: Is MSDS on file?_____*				
Spill contaminant system in place?				

HVAC Checklist - Long Form

Building: _____   File Number: _____

Completed by: _____   Title: _____   Date Checked: _____

Component	OK	Needs Attention	Not Applicable	Comments
Face and Bypass Dampers				
Damper operation correct?				
Damper motors operational?				
Cooling Coil				
Inspection access?				
Clean?				
Supply water temp. _____°F				
Water carryover?				
Any indication of condensation problems?				
Condensate Drip Pans				
Accessible to inspect and clean?				
Clean, no residue?				
No standing water, no leaks?				
Noticeable odor?				
Visible growth (e.g., slime)?				
Drains and traps clear, working?				
Trapped to air gap?				
Water overflow?				

HVAC Checklist - Long Form

Building: _____ File Number: _____

Completed by: _____ Title: _____ Date Checked: _____

Component	OK	Needs Attention	Not Applicable	Comments
Mist Eliminators				
Clean, straight, no carryover?				
Supply Fan Chambers				
Clean?				
No trash or storage?				
Floor drain traps are wet or sealed?				
No air leaks?				
Doors close tightly?				
Supply Fans				
Location ———————————				
Fan blades clean?				
Belt guards installed?				
Proper belt tension?				
Excess vibration?				
Corrosion problems?				
Controls operational, calibrated?				

HVAC Checklist - Long Form

Building: _____ File Number: _____

Completed by: _____ Title: _____ Date Checked: _____

Component	OK	Needs Attention	Not Applicable	Comments
Control sequence conforms to design/specifications? (describe changes)				
No pneumatic leaks?				
Heating Coil				
Inspection access?				
Clean?				
Control sequence conforms to design/specifications? (describe changes)				
Supply water temp. ____°F				
Discharge thermostat? (air temp. setting ____°F)				
Reheat Coils				
Clean?				
Obstructed?				
Operational?				
Steam Humidifier				
Humidifier type _____				
Treated boiler water?				
Standing water?				

HVAC Checklist - Long Form

Building: _____ File Number: _____

Completed by: _____ Title: _____ Date Checked: _____

Component	OK	Needs Attention	Not Applicable	Comments
Visible growth?				
Mineral deposits?				
Control setpoint _____°F				
High limit setpoint _____°F				
Duct liner within 12 feet? (If so, check for dirt, mold growth.)				
Supply Ductwork				
Clean?				
Sealed, no leaks, tight connections?				
Fire dampers open?				
Access doors closed?				
Lined ducts?				
Flex duct connected, no tears?				
Light troffer supply?				
Balanced within 3-5 years?				
Balanced after recent renovations?				
Short circuiting or other air distribution problems? Note location(s) _____ _____				
Pressurized Ceiling Supply Plenum				
No unintentional openings?				
All ceiling tiles in place?				

HVAC Checklist - Long Form

Building: _____ File Number: _____

Completed by: _____ Title: _____ Date Checked: _____

Component	OK	Needs Attention	Not Applicable	Comments
Barrier paper correctly placed and in good condition?				
Proper layout for air distribution?				
Supply diffusers open?				
Supply diffusers balanced?				
Balancing capability?				
Noticeable flow of air?				
Short circuiting or other air distribution problems? *Note location(s) in"Comments"*				
Terminal Equipment (supply)				
Housing interiors clean and unobstructed?				
Controls working?				
Delivering rated volume?				
Balanced within 3-5 years?				
Filters in place?				
Condensate pans clean, drain freely?				
VAV Box				
Minimum stops _____ %				
Minimum outside air ____ % *(from page 2 of this form)*				
Minimum airflow _____ cfm				
Minimum outside air _____ cfm				

HVAC Checklist - Long Form

Building: _____ File Number: _____

Completed by: _____ Title: _____ Date Checked: _____

Component	OK	Needs Attention	Not Applicable	Comments
Supply setpoint _____ °F (summer) _____ °F (winter)				
Thermostats				
Type _____				
Properly located?				
Working?				
Setpoints _____ °F (summer) _____ °F (winter)				
Space temperature _____ °F				
Humidity Sensor				
Humidistat setpoints _____ % RH				
Dehumidistat setpoints _____ % RH				
Actual RH _____ %				
Room Partitions				
Gap allowing airflow at top?				
Gap allowing airflow at bottom?				
Supply and return each room?				

HVAC Checklist - Long Form

Building: _____ File Number: _____

Completed by: _____ Title: _____ Date Checked: _____

Component	OK	Needs Attention	Not Applicable	Comments
Stairwells				
Doors close and latch?				
No openings allowing uncontrolled airflow?				
Clean, dry?				
No noticeable odors?				
Return Air Plenum				
Tiles in place?				
No unintentional openings?				
Return grilles?				
Balancing capability?				
Noticeable flow of air?				
Transfer grilles?				
Fire dampers open?				
Ducted Returns				
Balanced within 3-5 years?				
Unobstructed grilles?				
Unobstructed return air path?				
Return Fan Chambers				
Clean and no trash or storage?				
No standing water?				
Floor drain traps are wet or sealed?				

HVAC Checklist - Long Form

Building: _____ File Number: _____

Completed by: _____ Title: _____ Date Checked: _____

Component	OK	Needs Attention	Not Applicable	Comments
No air leaks?				
Doors close tightly, kept closed?				
Return Fans				
Location _____				
Fan blades clean?				
Belt guards installed?				
Proper belt tension?				
Excess vibration?				
Corrosion problems?				
Controls working, calibrated?				
Control sequence conforms to design/specifications? (describe changes)				
Exhaust Fans				
Central?				
Distributed (locations) _____				
Operational?				
Controls operational?				
Toilet exhaust only?				
Gravity relief?				

HVAC Checklist - Long Form

Building: _____ File Number: _____

Completed by: _____ Title: _____ Date Checked: _____

Component	OK	Needs Attention	Not Applicable	Comments
Total powered exhaust _____ cfm				
Make-up air sufficient?				
Toilet Exhausts				
Fans working occupied hours?				
Registers open, clear?				
Make-up air path adequate?				
Volume according to code?				
Floor drain traps wet or sealable?				
Bathrooms run slightly negative relative to building?				
Smoking Lounge Exhaust				
Room runs negative relative to building?				
Print Room Exhaust				
Room runs negative relative to building?				
Garage Ventilation				
Operates according to codes?				
Fans, controls, dampers all operate?				

HVAC Checklist - Long Form

Building: _____ File Number: _____

Completed by: _____ Title: _____ Date Checked: _____

Component	OK	Needs Attention	Not Applicable	Comments
Garage slightly negative relative to building?				
Doors to building close tightly?				
Vestibule entrance to building from garage?				
Mechanical Rooms				
General condition?				
Controls operational?				
Pneumatic controls:				
■ compressor operational?				
■ air dryer operational?				
Electric controls? Operational?				
EMS (Energy Management System) or DDC (Direct Digital Control):				
■ operator on site?				
■ controlled off-site?				
■ are fans cycled "off" while building is occupied?				
■ is chiller reset to shed load?				
Preventive Maintenance				
Spare parts inventoried?				
Spare air filters?				
Control drawing posted?				

HVAC Checklist - Long Form

Building: _____   File Number: _____

Completed by: _____   Title: _____   Date Checked: _____

Component	OK	Needs Attention	Not Applicable	Comments
PM (Preventive Maintenance) schedule available?				
PM followed?				
Boilers				
Flues, breeching tight?				
Purge cycle working?				
Door gaskets tight?				
Fuel system tight, no leaks?				
Combustion air: at least 1 square inch free area per 2000 Btu input?				
Cooling Tower				
Sump clean?				
No leaks, no overflow?				
Eliminators working, no carryover?				
No slime or algae?				
Biocide treatment working?				
Dirt separator working?				
Chillers				
No refrigerant leaks?				
Purge cycle normal?				
Waste oil, refrigerant properly disposed of and spare refrigerant properly stored?				
Condensation problems?				

HVAC Checklist - Long Form

Page ___ of ___

Building: _____ File Number: _____

Completed by: _____ Title: _____ Date Checked: _____

Component	OK	Needs Attention	Not Applicable	Comments

Pollutant Pathway Form For Investigations

Building Name: _____   File Number: _____

Address: _____   Completed by: _____

This form should be used in combination with a floor plan such as a fire evacuation plan.

Building areas that appear isolated from each other may be connected by airflow passages such as air distribution zones, utility tunnels or chases, party walls, spaces above suspended ceilings (whether or not those spaces are serving as air plenums), elevator shafts, and crawl spaces.

Describe the complaint area in the space below and mark it on your floor plan. Then list rooms or zones connected to the complaint area by airflow pathways. Use the form to record the direction of air flow between the complaint area and the connected rooms/zones, including the date and time. (Airflow patterns generally change over time). Mark the floor plan with arrows or plus (+) and minus (-) signs to map out the airflow patterns you observe, using chemical smoke or a micromanometer. The "Comments" column can be used to note pollutant sources that merit further attention.

Rooms or zones included in the complaint area: _____

_____

Sections 2, 4 and 6 discuss pollutant pathways and driving forces.

Rooms or Zones Connected to the Complaint Area By Pathways	Use	Pressure Relative to Complaint Area		Comments (e.g., potential pollutant sources)
		+/-	date/time	

Pollutant and Source Inventory

Building Name: _____ Address: _____

Completed by: _____ Date: _____ File Number: _____

Using the list of potential source categories below, record any indications of contamination or suspected pollutants that may require further investigation or treatment. Sources of contamination may be constant or intermittent or may be linked to single, unrepeated events. For intermittent sources, try to indicate the time of peak activity or contaminant production, including correlations with weather (e.g., wind direction).

Sections 2, 4 and 6 discuss pollutant sources. Appendix A provides guidance on common measurements.

Source Category	Checked	Needs Attention	Location	Comments
SOURCES OUTSIDE BUILDING				
Contaminated Outdoor Air				
Pollen, dust				
Industrial contaminants				
General vehicular contaminants				
Emissions from Nearby Sources				
Vehicle exhaust (parking areas, loading docks, roads)				
Dumpsters				
Re-entrained exhaust				
Debris near outside air intake				
Soil Gas				
Radon				
Leaking underground tanks				
Sewage smells				
Pesticides				

Pollutant and Source Inventory

Building Name: _____ Address: _____

Completed by: _____ Date: _____ File Number: _____

Using the list of potential source categories below, record any indications of contamination or suspected pollutants that may require further investigation or treatment. Sources of contamination may be constant or intermittent or may be linked to single, unrepeated events. For intermittent sources, try to indicate the time of peak activity or contaminant production, including correlations with weather (e.g., wind direction).

Source Category	Checked	Needs Attention	Location	Comments
Moisture or Standing Water				
Rooftop				
Crawlspace				
EQUIPMENT				
HVAC System Equipment				
Combustion gases				
Dust, dirt, or microbial growth in ducts				
Microbial growth in drip pans, chillers, humidifiers				
Leaks of treated boiler water				
Non HVAC System Equipment				
Office Equipment				
Supplies for Equipment				
Laboratory Equipment				

Pollutant and Source Inventory

Building Name: _____ Address: _____

Completed by: _____ Date: _____ File Number: _____

Using the list of potential source categories below, record any indications of contamination or suspected pollutants that may require further investigation or treatment. Sources of contamination may be constant or intermittent or may be linked to single, unrepeated events. For intermittent sources, try to indicate the time of peak activity or contaminant production, including correlations with weather (e.g., wind direction).

Source Category	Checked	Needs Attention	Location	Comments
HUMAN ACTIVITIES				
Personal Activities				
Smoking				
Cosmetics (odors)				
Housekeeping Activities				
Cleaning materials				
Cleaning procedures (e.g., dust from sweeping, vacuuming)				
Stored supplies				
Stored refuse				
Maintenance Activities				
Use of materials with volatile compounds (e.g., paint, caulk, adhesives)				
Stored supplies with volatile compounds				
Use of pesticides				

Pollutant and Source Inventory

Building Name: _____ Address: _____

Completed by: _____ Date: _____ File Number: _____

Using the list of potential source categories below, record any indications of contamination or suspected pollutants that may require further investigation or treatment. Sources of contamination may be constant or intermittent or may be linked to single, unrepeated events. For intermittent sources, try to indicate the time of peak activity or contaminant production, including correlations with weather (e.g., wind direction).

Source Category	Checked	Needs Attention	Location	Comments
BUILDING COMPONENTS FURNISHINGS				
Locations Associated with Dust or Fibers				
Dust-catching area (e.g., open shelving)				
Deteriorated furnishings				
Asbestos-containing materials				
Unsanitary Conditions/Water Damage				
Microbial growth in or on soiled or water-damaged furnishings				

Pollutant and Source Inventory

Building Name: _____ Address: _____

Completed by: _____ Date: _____ File Number: _____

Using the list of potential source categories below, record any indications of contamination or suspected pollutants that may require further investigation or treatment. Sources of contamination may be constant or intermittent or may be linked to single, unrepeated events. For intermittent sources, try to indicate the time of peak activity or contaminant production, including correlations with weather (e.g., wind direction).

Source Category	Checked	Needs Attention	Location	Comments
Chemicals Released From Building Components or Furnishings				
Volatile compounds				
OTHER SOURCES				
Accidental Events				
Spills (e.g., water, chemicals, beverages)				
Water leaks or flooding				
Fire damage				

Pollutant and Source Inventory

Page 6 of 6

Building Name: _____ Address: _____

Completed by: _____ Date: _____ File Number: _____

Using the list of potential source categories below, record any indications of contamination or suspected pollutants that may require further investigation or treatment. Sources of contamination may be constant or intermittent or may be linked to single, unrepeated events. For intermittent sources, try to indicate the time of peak activity or contaminant production, including correlations with weather (e.g., wind direction).

Source Category	Checked	Needs Attention	Location	Comments
Special Use/Mixed Use Areas				
Smoking lounges				
Food preparation areas				
Underground or attached parking garages				
Laboratories				
Print shops, art rooms				
Exercise rooms				
Beauty salons				
Redecorating/Repair/Remodeling				
Emissions from new furnishings				
Dust, fibers from demolition				
Odors, volatile compounds				

Pollutant and Source Inventory

Page___of___

Building Name: _____ Address: _____

Completed by: _____ Date: _____ File Number: _____

Using the list of potential source categories below, record any indications of contamination or suspected pollutants that may require further investigation or treatment. Sources of contamination may be constant or intermittent or may be linked to single, unrepeated events. For intermittent sources, try to indicate the time of peak activity or contaminant production, including correlations with weather (e.g., wind direction).

Sections 2, 4 and 6 discuss pollutant sources. Appendix A provides guidance on common measurements.

Source Category	Checked	Needs Attention	Location	Comments

Chemical Inventory

Building Name: _____ File Number: _____

Address: _____

Completed by: _____ Phone: _____

The inventory should include chemicals stored or used in the building for cleaning, maintenance, operations, and pest control. If you have an MSDS (Material Safety Data Sheet) for the chemical, put a check mark in the right-hand column. If not, ask the chemical supplier to provide the MSDS, if one is available.

Sections 2,4 and 6 discuss pollutant sources. Section 4 discusses MSDSs.

Date	Chemical/Brand Name	Use	Storage Location(s)	MSDS on file?

Hypothesis Form

Page 1 of 2

Building Name: _____ File Number: _____

Address: _____

Completed by: _____

Complaint Area (may be revised as the investigation progresses):

_____

_____

_____

_____

Complaints (e.g., summarize patterns of timing, location, number of people affected):

_____

_____

HVAC: Does the ventilation system appear to provide adequate outdoor air, efficiently distributed to meet occupant needs in the complaint area? If not, what problems do you see?

_____

_____

Is there any apparent pattern connecting the location and timing of complaints with the HVAC system layout, condition or operating schedule?

_____

_____

Pathways: What pathways and driving forces connect the complaint area to locations of potential sources?

_____

Are the flows opposite to those intended in the design? _____

Sources: What potential sources have been identified in the complaint area or in locations associated with the complaint area (connected by pathways)?

_____

Is the pattern of complaints consistent with any of these sources? _____

Hypothesis Form

Hypothesis: Using the information you have gathered, what is your best explanation for the problem?

_____

_____

_____

Hypothesis testing: How can this hypothesis be tested?

_____

_____

_____

_____

If measurements have been taken, are the measurement results consistent with this hypothesis?

_____

_____

_____

Results of Hypothesis Testing:

_____

_____

_____

_____

Additional Information Needed:

_____

_____

_____

_____

Index